Infrared Design Examples

TUTORIAL TEXTS SERIES

Infrared Design Examples

William L. Wolfe

Professor Emeritus
Optical Sciences Center, University of Arizona

Tutorial Texts in Optical Engineering
Volume TT36

Donald C. O'Shea, Series Editor
Georgia Institute of Technology

SPIE OPTICAL ENGINEERING PRESS
A Publication of SPIE—The International Society for Optical Engineering
Bellingham, Washington USA

Library of Congress Cataloging-in-Publication Data

Wolfe, William L.
    Infrared design examples / William L. Wolfe
      p. cm. – (Tutorial texts in optical engineering; v. TT36)
    Includes bibliographical references and index.
    ISBN 0-8194-3319-5 (softcover)
    1. Infrared equipment–Design and construction I. Series.
TA1570.W64 1999
621.36'2—dc21   99-36329
                                            CIP

Published by

SPIE—The International Society for Optical Engineering
P.O. Box 10
Bellingham, Washington 98227-0010
Phone: 360/676-3290
Fax: 360/647-1445
Email: spie@spie.org
WWW: http://www.spie.org/

Printed in the United States of America.

Second Printing

SERIES INTRODUCTION

The Tutorial Texts series was begun in response to requests for copies of SPIE short course notes by those who were not able to attend a course. By policy the notes are the property of the instructors and are not available for sale. Since short course notes are intended only to guide the discussion, supplement the presentation, and relieve the lecturer of generating complicated graphics on the spot, they cannot substitute for a text. As one who has evaluated many sets of course notes for possible use in this series, I have found that material unsupported by the lecture is not very useful. The notes provide more frustration than illumination.

What the Tutorial Texts series does is to fill in the gaps, establish the continuity, and clarify the arguments that can only be glimpsed in the notes. When topics are evaluated for this series, the paramount concern in determining whether to proceed with the project is whether it effectively addresses the basic concepts of the topic. Each manuscript is reviewed at the initial state when the material is in the form of notes and then later at the final draft. Always, the text is evaluated to ensure that it presents sufficient theory to build a basic understanding and then uses this understanding to give the reader a practical working knowledge of the topic. References are included as an essential part of each text for the reader requiring more in-depth study.

One advantage of the Tutorial Texts series is our ability to cover new fields as they are developing. In fields such as sensor fusion, morphological image processing, and digital compression techniques, the textbooks on these topics were limited or unavailable. Since 1989 the Tutorial Texts have provided an introduction to those seeking to understand these and other equally exciting technologies. We have expanded the series beyond topics covered by the short course program to encompass contributions from experts in their field who can write with authority and clarity at an introductory level. The emphasis is always on the tutorial nature of the text. It is my hope that over the next few years there will be as many additional titles with the quality and breadth of the first ten years.

Donald C. O'Shea
Georgia Institute of Technology

TABLE OF CONTENTS

PREFACE

This text is an extension of my earlier book, *Introduction to Infrared System Design*,[1] in which materials, detectors, optics, scanners, and sensitivity calculations were introduced. This text builds on the principles and information presented in the earlier work and addresses several problems in detail. Most of the problems are real, or closely resemble actual problems.

The MX Shell Game was proposed to me some years ago when the people who were considering housing the Peacekeeper wanted to know if the position of the vehicle could be detected by the "Red Team." I have altered it slightly to illustrate the principles of designing a demanding strip mapper. The final sections of Chapter 2 discuss less-demanding applications, some of which have been implemented. The Space Campout is a whimsical design of an instrument that detects ICBMs in midcourse. It illustrates principles that I have applied in addressing many varied design problems over the years.

The third problem, the night driving system, I first tackled at Honeywell for tanks, and most recently at General Motors for civilian cars. I never really looked at the BOSS, but my imagination was piqued by Tom Clancy's book,[2] and I considered the design of an infrared detection system that could track surface vessels on the oceans of the world in place of Clancy's radar surveillance device, the RORSAT. Again, similar problems exist in the scientific and corporate milieus. The collision avoidance system for airplanes was considered in the 80s for civilian applications, and it exists in the military as tail-warning systems that protect the aircraft from behind from a variety of missiles and aircraft attacks. The helicopter pilot system (described briefly in Chapter 5) was carried out for McDonnell Douglas in Mesa. The design presented here is mine, not necessarily the one chosen by them or the Army. It is a nice example of how vision and display considerations can, and should, determine many aspects of the design.

The applications are military, industrial, aeronautical, space, medical, and even piscatorial. Almost every military application has a civilian counterpart and *vice versa*. The applications are driven by different requirements, including high probability of detection, speed, cost, and size. They employ photon and thermal detectors, starers and scanners, pushbrooms and whiskbrooms, the midwave and longwave spectral bands, automobiles, satellites, airplanes, and ears.

I have not dealt with spectrometers, LIDARS, other active systems, or pollution monitoring. Maybe next time.

---

[1] W. L. Wolfe, *Introduction to Infrared System Design,* SPIE Press (1996).

[2] T. Clancy, *Red Storm Rising*, Putnam and Sons (1986).

I wrote this text in a personal way, because I dealt with these problems personally. I chose a range of problems that illustrate most of the design issues encountered in the field of infrared technology.

Each problem is discussed in almost a stream-of-consciousness genre, and that is intentional. I have tried to solve each problem as an engineer would. The classical techniques of many texts, in which an inexorable and direct approach to the proper solution is described, has not been used here. I don't think I have ever made all the right decisions at every step of the way in any design—or most other aspects of my life!

The design process outlined in Chapter 2 works in almost all cases. However, infrared design is something of an art based on experience. Just as a cheetah goes for the jugular, the experienced infrared-system designer must recognize at the outset the critical aspects of the design. This means that, although I have outlined a rational sequential procedure, there are times when one must jump around. The plodder will get there, but maybe only after a good deal of unnecessary effort. It is better to be smart and lazy. And it is even better to be lucky!

Every one of these systems involves electronics and mechanics. Details of these and other required, related disciplines have not been included so that the time and space could be used for the infrared aspects of the problems.

I chose an unorthodox approach in the presentation of the equations; they are Mathcad printouts. Many of the figures have a combination of these equations and plots of the data for some domain of an independent variable. For those who are not conversant with Mathcad, I have described the equations line by line, and repeated them in the body of the text. To minimize repetition, I assumed that the reader will become familiar with the presentation and I became much briefer. For those who are not conversant, I recommend starting with the MX shell game; it has the most detailed explanations of the Mathcad formalism.

To set a value for a variable, say, for example, the speed of light, the format is $c:=2.9975 \times 10^{10}$. The colon followed by the equals sign indicates a setting. The same is true for an equation $f:=m*a$. The asterisk signifies multiplication. When only an equals sign is used, then a value has been calculated: $v=7.4$. The domain of an independent variable is set with the first value, a comma, the next value, a semicolon, and the last value: $t:=1,2;50$. Dependent variables must have their associated independent variables in parentheses: $v(t):=s/t$. Both the independent and the dependent variables are included on the axes of graphs. Experienced users will recognize that there are many features of Mathcad that I have not employed for the sake of simplicity. One of these is the incorporation of units with the variables as something of a sanity check.

I would like to again acknowledge the editorial acumen of Don O'Shea, who provided many useful comments, as he did on my earlier Tutorial Texts. Bjorn Andresen carefully read and made many worthwhile and perspicacious corrections. A debt is owed to an anonymous reviewer, Bob Fischer, who also contributed valuable comments. My wife suffered me and wondered about the time I spent in front of the computer—but now she has one too. I may try another text. I hope these examples and the style in which they are presented are useful to the reader.

I dedicate this text to those who went before me and affected my career—Stan Ballard, Luc Biberman—mentors both; Kiyo Tomiyasu, Lloyd Mundie, Gwynn Suits, George Zissis, Mike Holter, Bob Rynearson, Aden Meinel, Peter Franken, Bob Shannon, and Dick Powell—bosses all. And all are—or were—friends.

William L. Wolfe
July 1999

Infrared Design Examples

CHAPTER 1
REVIEW OF DESIGN FUNDAMENTALS

This chapter reviews principles detailed in *Introduction to Infrared System Design*,[1] and recaps the important equations and concepts from that text. In addition, it outlines the procedures that I have found to be efficient when designing infrared systems.

1.1 THE DESIGN PROCESS

Infrared system design is not a synthetic process, as is some circuit design. One cannot proceed in an orderly fashion from the statement of the problem to the final solution. Rather, it is necessary to "guess" a solution and explore its relevance. Infrared system design is iterative. The iteration goes faster, as with most iterations, if the first guess is good. The quality of that first guess is surely a function of insight and experience, and--I guess--insight follows experience.

The infrared-system design process that I have found to be most efficient is to calculate, in order, the geometry, the dynamics, the sensitivity, and, finally, the optics. Then it is important to review the results, consider the alternatives, and try again.

The geometry may be large or small, astronomical or microscopic. It may be in the sky, in space, or on the ground, but the essential considerations include the angular resolution, total angular field of view, the number of pixels in the field, and the range to the center and the edges of the field.

The dynamics include the frequency with which the field must be scanned (the frame time), the line time and dwell time, and, therefore, the required bandwidth. The bandwidth is required for the next step--the sensitivity calculation.

In the case of an imager, the sensitivity calculation is usually the minimum resolvable temperature difference (MRTD), which is closely related to the noise-equivalent temperature difference (NETD). For detection systems, sensitivity is usually indicated by the signal-to-noise ratio, or it may be the resultant probability of detection and false alarm rate. The calculation at this stage can be idealized, perhaps assuming that all efficiency factors are 100%. Because at least one more iteration will occur, precision is not necessary. In fact, the approximate, simplified calculations will aid in reducing the solution space. The radiometric evaluation of the target and background are an inherent part of the sensitivity analysis. Approximations can be relatively loose in the first iteration, but they must be refined later. After some refining, the optics and detector size can be approximated. Guesses at the spectral bands can be made. Alternatively, the entire spectrum can be used, according to the Stefan Boltzmann law.

1

The next step is to approximate the optical system. I have applied gaussian optics and the third-order blur equations, Eqs. (1.8) through (1.11), for many years. The diffraction limit should be evaluated, because you can't do better than that. A detailed optical design will generate an optical system that performs better than the blur equations predict (see Appendix B). Consideration of the optics requirements often must include consideration of scanning requirements.

Figure 1.1 illustrates the design process. One can insert the efficiencies as soon as they are known, but they are generally not known until a system has been postulated. These four steps–geometry, dynamics, sensitivity, and evaluation of the

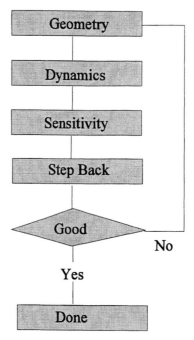

Figure 1.1. The design process

optics--are repeated several times to achieve a near-optimum spectral region, altitude, range, optics diameter, optical system, or scan technique.

Then one should consider and calculate the various efficiencies and their effects on the performance. When considering photon-limited detectors, the assumption should be checked. The optical efficiency should be calculated, as should the atmospheric transmission. Assumptions about the source might need to be refined.

An essential step to the best solution is a step back. Are other approaches--a different altitude, different spectral region or detectors, even multiple optical systems–a more appropriate solution? This step back can be the innovative step that wins a contract or generates a patent.

1.2 GEOMETRY

The angular pixel or resolution element can usually be calculated as a simple ratio. It is usually square and is given simply by the ratio of the side of the pixel to the detection range. When square, most designers state only the one angle. The symbols used here are a and b for the linear dimensions of a detector element and α and β for the angular dimensions. The field of view is usually much larger and must be calculated with

$$\Theta = 2\arctan\left(\frac{S}{2R}\right),$$
(1.1)

where Θ is the full field of view, S is the full length or width of the field, and R is the perpendicular range from the sensor to the center of the field. A similar equation applies to the other dimension of the field if the field is not square. A large angular field complicates the calculation of the angular resolution. Figure 1.2 shows that the range at the edge of the field is larger than the perpendicular range, and a projection is necessary.

The angular pixel size is given by the two equations for the sides

$$\alpha = \frac{x\cos\theta}{R} \quad \beta = \frac{y\cos^2\theta}{R},$$
(1.2)

where α and β are the angular measures of the resolution element, x and y are the measures of the resolution on the object, R is the slant range, and θ is the full-field angle. To calculate the corner edge of the field, the Pythagorean theorem must be applied.

1.3 DYNAMICS

For real-time imagers, the frame rate is 1/30th of a second, the U.S. television frame rate. For interceptors, it is dictated by closing rates or search times, and for strip mappers, it is given by the velocity-to-height ratio and resolution. The dwell time is the frame time divided by the number of pixels and an estimated scan efficiency and multiplied by the number of detectors used. The bandwidth is then one divided by twice the dwell time. That is

$$B = \frac{1}{2\eta_{sc}t_f}\frac{N_h N_v}{m_h m_v} \, , \tag{1.3}$$

where N represents the number of angular pixels in the field in the horizontal and vertical directions and m represents the number of detector elements in those directions, η_{sc} is the scan efficiency, and t_f is the frame time.

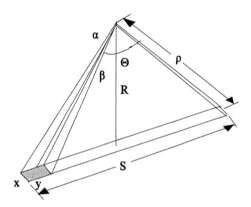

Figure 1.2. Geometry.

1.4 SENSITIVITY

The sensitivity equations were developed *ad nauseam* in *Infrared System Design* (SPIE Tutorial Text TT24). The most useful ones will be repeated here without derivation.

The signal-to-noise ratio (SNR), for a point source in terms of a specific detectivity is

$$SNR = \frac{D^*\Phi_d}{\sqrt{A_d B}} = \frac{\bar{\tau}_a \bar{\tau}_o \bar{\varepsilon} A_s A_o}{R^2 \sqrt{A_d B}} \int D^* L_\lambda^{BB} d\lambda, \tag{1.4}$$

where $D*$ is the specific detectivity, Φ_d is the power on the detector from the source, B is the effective noise bandwidth, τ represents the transmittances of the atmosphere and the optics, ε is the emissivity of the source, A_d, A_s, and A_o represent detector, source, and optical aperture areas, R is the range, and L_λ^{BB} is the spectral radiance of the blackbody source. The overbars indicate weighted average values, as described in *Infrared System Design*.[1] If the detector is limited by photon noise and the optics by the diffraction limit, then

$$SNR = \bar{\varepsilon}\bar{\tau}_a\bar{\tau}_o\sqrt{\eta\eta_{cs}}\frac{A_sA_o}{R^2\sqrt{A_dB}}\frac{\int L_q\lambda d\lambda}{\int E_q\lambda d\lambda}, \qquad (1.5)$$

where η represents the detector quantum efficiency and the cold-shielding efficiency, $E_{q\lambda}$ is the photon incidance, and photonic quantities are represented by the subscript q.

Most extended-source infrared applications deal with NETDs and the two equations that provide this information are

$$NETD = \frac{4}{\pi}\frac{F}{D_o}\frac{\sqrt{B}}{\bar{\tau}_a\bar{\tau}_o\varepsilon}\frac{1}{\alpha}\frac{1}{D*dL^{BB}/dT} \qquad (1.6)$$

and

$$NETD_{BLISS} = 1.35\frac{\sqrt{B}}{\lambda\bar{\tau}_a\bar{\tau}_o\varepsilon}\sqrt{\frac{g}{\eta\eta_{cs}}\frac{\sqrt{L_q}}{dL_q^{BB}/dT}}, \qquad (1.7)$$

where g is either 4 or 2, depending on the absence or presence of recombination or carriers. Equation (1.7) is for a photon-limited, diffraction-limited system.

1.5 OPTICS

Assuming that things have not fallen apart in the previous steps, apply the approximate third-order equations to look for reasonable optics solutions and then calculate the diffraction limit. The approximate equations for the angular blur diameters based on the diffraction limit, spherical aberration, coma, and astigmatism are, respectively,

$$\beta_{DL} = \frac{2.44\lambda}{D_o} \tag{1.8}$$

$$\beta_{SA} = \frac{1}{128F^3} \tag{1.9}$$

$$\beta_{CA} = \frac{\Theta}{16F^2} \tag{1.10}$$

$$\beta_{AA} = \frac{\Theta^2}{2F}. \tag{1.11}$$

The diffraction limit, as presented here, is related to the Rayleigh limit, but actually represents the diameter of the main lobe of the diffraction pattern of a circular aperture. The detector should just fill this blur.[2] Several designs are available in the following sources: for mirror and catadioptric systems, the article by Jones[3] is useful. Most good design programs have libraries of existing optical designs. I have included the main figure from Jones' article and descriptions of the most applicable designs in Appendix B.

1.6 STEP BACK

This is the time to consider all reasonable alternatives--a different detector, a different platform, other spectral regions, different optics, beamsplitters, and multiplexers in field, aperture, time, space, or--even-- profession. Step back.

1.7 ITERATE

In performing the steps above you should have learned something. Use the information wisely. Change or keep the spectral range, the frame time, the detector, the range, the optics size, or anything else you may. Or (the Good Lord willin') proceed to refine the calculations.

The steps described somewhat generally here will become more meaningful as applied to the problems addressed in the chapters that follow.

1.8 REITERATE
1.8 REITERATE
1.8 REITERATE
1.8 REITERATE
1.8 REITERATE
1.8 REITERATE
1.8 REITERATE
1.8 REITERATE
1.8 REITERATE

1.9 REFERENCES

[1] W. L. Wolfe, *Introduction to Infrared System Design*, SPIE (1996).

[2] R. B. Emmons, personal communication.

[3] L. Jones,, "Reflective and catadioptric objectives," in M. Bass, E. Van Stryland, D. Williams, and W. Wolfe, eds., *Handbook of Optics*, McGraw Hill, 1995.

CHAPTER 2
THE MX SHELL GAME

During the height of the Cold War, it was proposed that Peacekeeper missiles be housed in tunnels in the Southwest. I hoped that meant California, Nevada, or Utah–and not Arizona. The missiles would be shuffled around in underground tunnels so that they could not be pinpointed for destruction by incoming missiles. The proponents of this plan wondered whether there was a way to detect the position of the missiles in the tunnels by a remote method. My investigation of an infrared detection system is the basis for this chapter.

2.1 PROBLEM STATEMENT

The area is 200 km by 200 km and located in Nevada. The 20-km-long tunnels are oriented and placed at random. It is assumed that a 10-m-wide spot above the missile has a temperature that is 1 K above the surrounding surface temperature because the missile must be kept in readiness and therefore has associated operating machinery (Fig. 2.1). "Your problem, Mr. Phelps," as they said on the old television show, *Mission Impossible*, "is to determine if you can find this spot."

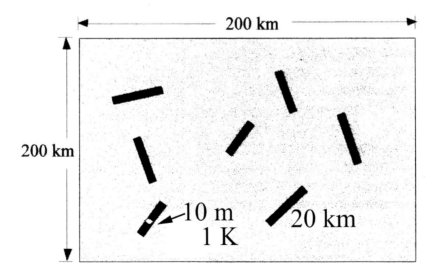

Figure 2.1. Illustration of the Peacekeeper tunnel layout and a 10-m spot, 1 K above ambient.

In the usual context of multiple choice exams, you must answer the following questions:

1. At what altitude should you fly?
    a. Geosynchronous orbit
    b. High-altitude airplane altitude
    c. Low-altitude orbit
    d. All of the above
2. What spectral region should you use?
    a. The visible
    b. Midwave infrared, 3-5 µm
    c. Long-wave Infrared, 8--12 µm
    d. Everything Messers Stefan and Boltzmann allow
3. What kind of scanner should you use?
    a. Pushbroom
    b. Whiskbroom
    c. Mop
    d. Step starer
4. What kind of detectors should you use?
    a. Photon
    b. Thermal
    c. Pyroelectric
    d. Extrinsic
    e. Intrinsic
    f. ESP

To answer these questions without using a dart, we must start the design process--geometry, dynamics, sensitivity, optics, step back, and iteration.

2.2 GEOMETRY

This unit can be flown at three possible altitudes: geosynchronous orbit, low-altitude earth orbit, and high aircraft altitude.

Geosynchronous orbit is 33,000 km, or 33 Mm, above the Earth (see Fig. 2.2). For a geosynchronous orbit, the 10-m ground pixel subtends 0.3 µrad. The half-field angle is 3 mrad. The very small angular subtense of the ground pixel should raise the hackles on the neck of any seasoned infraredder. The diffraction-limited diameter for a 12-µm system is

$$D = \frac{2.44\lambda}{\alpha} = \frac{2.44 x 12 x 10^{-6}}{0.3 x 10^{-6}} = 8.13m. \tag{2.1}$$

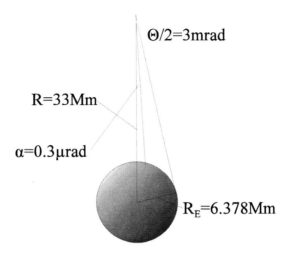

Figure 2.2. Geosynchronous-orbit geometry.

For the midwave infrared region (MWIR), the mirror must be 3.38 m (which is huge), and even the diameter for the visible system will be 34 cm (and a visible system will not sense the temperature differences). Thus, we cannot use a geosynchronous orbit, and we cannot use the visible.

A high-flying aircraft (at 33 km or about 100,000 ft) has different problems. Figure 2.3 shows the geometry. The half field is calculated as

$$\Theta = \arctan\left(\frac{S}{2R}\right) = \arctan\left(\frac{200}{2x33}\right) = 1.252 rad = 71.7°. \qquad (2.2)$$

The full field is 143.4 degrees (very large). The angular pixel size must be calculated using the cosine-squared formalism:

$$\alpha = \frac{x}{\rho}\cos(\theta) = \frac{x}{R}\cos^2(\frac{\Theta}{2}) = \frac{0.01}{33}\cos^2(1.252) = 30\mu rad. \qquad (2.3)$$

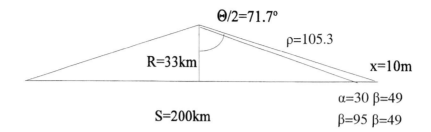

Figure 2.3. High-aircraft geometry.

This yields microradian resolution across a full field of almost 150 degrees, which will not work unless the field is reduced and the plane makes a number of passes. Such a plane will not last long; it might fly up from Mexico and make four or five passes in Nevada at perhaps 500 mph then and return to Mexico or to an aircraft carrier in the Sea of Cortez.

The only practical approach is to use a satellite as a platform in a low-altitude Earth orbit, between 100 and 1000 km. Figure 2.4 shows the angular field, the angular resolution, and the number of detectors required for a pushbroom system. (The number of detectors also indicates the degree of expense and difficulty for any system.) The flattening of the field-of-view curves lead one to choose altitudes at or near 400 km. This is supported by the angular-resolution and detector-number curves in the following way. The half-field of view decreases from about 45 degrees at 100 km (90 degrees, or 1.6 rad full field) to about 14 degrees (0.5 rad full field at 400 km). The required resolution decreases from about 70 µrad at 100 km to about 22 µrad at 400 km. The field decreased by a factor of about 3, while the required resolution decreased by about the same factor. However, the astigmatism is a quadratic factor. The proof of the pudding is in the final calculations, but this is good at the recipe stage.

Figure 2.4 is the first of many figures in this text based on Mathcad calculations. Some explanation is in order. The first equation, R:=100, 101..1000 indicates that the calculations are a function of the independent variable R, and that the variable will run from a minimum value of 100 to a maximum of 1000 in steps of 1, i.e., 100, then 101, to 1000:

$$R := 100, 101..1000 \; . \qquad\qquad (2.4)$$

In Mathcad, the colon followed by the equals sign indicates a defining equation.

$R := 100, 101 .. 1000$ $S := 200$ $x := 0.01$ $\Theta(R) := atan\left(\dfrac{S}{2 \cdot R}\right)$ $\alpha(R) := \dfrac{x \cdot cos(\Theta(R))}{R} \cdot 10^6$

$\beta(R) := \alpha(R) \cdot cos(\Theta(R))$ $N(R) := \dfrac{2 \cdot \Theta(R)}{\beta(R)} \cdot 10^6$ $\Theta(200) = 0.464$

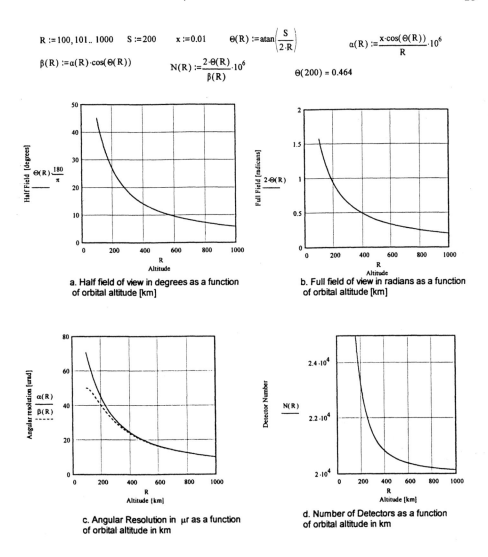

a. Half field of view in degrees as a function
of orbital altitude [km]

b. Full field of view in radians as a function
of orbital altitude [km]

c. Angular Resolution in μr as a function
of orbital altitude in km

d. Number of Detectors as a function
of orbital altitude in km

Figure 2.4. a)Half field of view as a function of orbital altitude (km); b)Full field of view in radians as a function of orbital altitude (km); c) angular resolution in μr as a function of orbital altitude in km; d) number of detectors as a function of orbital altitude.

The simple equals sign obtains the value for a given function or variable. Thus, equation (2.5) sets the value of the swath width at 200 km:

$$S := 200 .$$ (2.5)

The ground resolution at 10 m, 0.01 km, is specified by

$$x := 0.01 .$$
(2.6)

Using Eq. (1.1), we can calculate the half angle of the full field:

$$\Theta(R) := atan\left(\frac{S}{2*R}\right) .$$
(2.7)

In Mathcad, the function atan(x) means the arctangent of x, or the inverse tangent of x. Equation (2.8) calculates the resolution angle in the cross-scan direction. The penultimate equation calculates the in-scan resolution angle:

$$N(R) = \frac{2*\Theta(R)}{\beta(R)} * 10^6 .$$
(2.8)

Equation (2.8) gives the number of resolution elements, or the required number of detectors for a pushbroom, using the factor of a million to convert from microradians to radians. Then, the simple equation, with no colon, gives the value of the field of view at 200 km in radians:

$$\Theta(200) = 0.464 .$$
(2.9)

We will continue by assuming that the system uses a pushbroom scanner. It will be followed by analysis of the whisk broom and the step stare. One reason for this sequence is that the pushbroom has no moving parts and would be convenient to implement. The pushbroom geometry is shown in Fig. 2.5. Using Eqs. (1.1) and (1.2), we find that the half field of view is 0.2 rad, the full field 0.4 rad, and the angular subtenses are 9.8 and 8.6 μrad, for an altitude of 1 Mm. We will reassess this in Section 2.5 on optics.

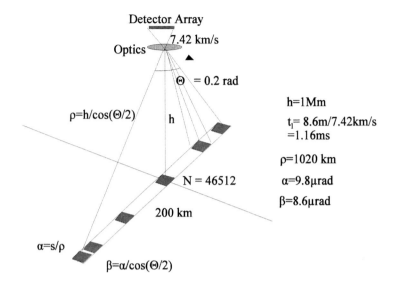

Figure 2.5 Pushbroom geometry.

2.3 DYNAMICS

A satellite at a 1-Mm altitude travels at about 7.42 km/s. This can be determined from tables, but the following short calculation supports it as well. The orbital time is approximately 90 min. The circumference of the Earth is approximately 40,000 km. Division of the distance by the time in seconds yields a value of 7.40. The satellite is at an altitude that ranges from 100 km to 1000 km, thereby increasing to distances that range from 40,700 to 46,300 and velocities of to 7.42 to 7.59. Appendix C gives a more accurate formula and some data. Taking the pixel at nadir (vertically below the vehicle) as 10 m, the dwell time for the pushbroom is 1.3 ms.

2.4 SENSITIVITY

The sensitivity is properly calculated as the idealized NETD. I have translated the 1-K temperature difference requirement to an NETD requirement of 0.1 K. If we attain this NETD, we will have a signal-to-noise ratio (SNR) of 10--a good value. Whether a SNR of 10 is adequate is open to discussion. In Chapter 6, detection and false alarm rates are considered, but only for non-imaging devices. Imagery provides much information for the eye and brain to use. From my experience, a SNR of 10 is a good value for imagery. The NETD may be calculated in an idealistic way as

$$L_q = \int_{\lambda_1}^{\lambda_2} \frac{2c}{\lambda^4(e^{c_2/\lambda T}-1)} d\lambda , \qquad (2.11)$$

where the photance (photonic radiance) is given by

$$NETD = \frac{1.35\sqrt{L_q B}}{\lambda_{ave} dL_q} , \qquad (2.10)$$

the bandwidth, B, was calculated above, the average wavelength λ_{ave} is

$$\lambda_{ave} = \frac{\lambda_1 + \lambda_2}{2} , \qquad (2.12)$$

and dL_q is the change in photance with respect to temperature,

$$dL_q = \frac{\partial L_q}{\partial T} = \int_{\lambda_1}^{\lambda_2} \frac{2c\frac{c_2}{\lambda T}e^{c_2/\lambda T}}{\lambda^4(e^{c_2/\lambda T}-1)^2 T} d\lambda . \qquad (2.13)$$

The calculation is shown in Fig. 2.6 for the 3- to 5-µm range and the 8- to 12-µm range and for the dwell time calculated above. The idealized equations are used with all efficiencies set to one. This is a good first calculation, partly because all the efficiencies are not yet known. Note, that in these calculations, the wavelengths are given in centimeters. I have found this to be convenient. Then the velocity of light is approximately 3×10^{10} centimeters per second and the second radiation constant is 1.4388 cm K.

     In Fig. 2.6, which is a Mathcad calculation, the first line sets the temperature to 300 K, the short wavelength of the spectral band to 3 µm (0.0003 cm), the long-wave limit to 5 µm (0.00005 cm), the second radiation constant to 1.4388, and the speed of light to 2.9975×10^{10} cm/s.

$T := 300$ $\lambda_1 := 0.0003$ $\lambda_2 := 0.0005$ $c_2 := 1.4388$ $c := 2.9975 \cdot 10^{10}$

$$B := \frac{1}{2 \cdot 0.001}$$

$$L_q := \int_{\lambda_1}^{\lambda_2} \frac{2 \cdot c}{\lambda^4 \cdot \left(e^{\frac{c_2}{\lambda \cdot T}} - 1\right)} \, d\lambda$$

$$dL_q := \int_{\lambda_1}^{\lambda_2} \frac{2 \cdot c \cdot \dfrac{c_2}{\lambda \cdot T} \cdot e^{\frac{c_2}{\lambda \cdot T}}}{\lambda^4 \cdot \left(e^{\frac{c_2}{\lambda \cdot T}} - 1\right)^2 \cdot T} \, d\lambda$$

$$\lambda_{ave} := \frac{\lambda_1 + \lambda_2}{2}$$

$$NETD := \frac{1.35 \cdot \sqrt{L_q \cdot B}}{\lambda_{ave} \cdot dL_q}$$

$$NETD = 0.033$$

$\lambda_1 := 0.0008$ $\lambda_2 := 0.0012$

$$L_q := \int_{\lambda_1}^{\lambda_2} \frac{2 \cdot c}{\lambda^4 \cdot \left(e^{\frac{c_2}{\lambda \cdot T}} - 1\right)} \, d\lambda$$

$$dL_q := \int_{\lambda_1}^{\lambda_2} \frac{2 \cdot c \cdot \dfrac{c_2}{\lambda \cdot T} \cdot e^{\frac{c_2}{\lambda \cdot T}}}{\lambda^4 \cdot \left(e^{\frac{c_2}{\lambda \cdot T}} - 1\right)^2 \cdot T} \, d\lambda$$

$$\lambda_{ave} := \frac{\lambda_1 + \lambda_2}{2}$$

$$NETD := \frac{1.35 \cdot \sqrt{L_q \cdot B}}{\lambda_{ave} \cdot dL_q}$$

$$NETD = 4.249 \cdot 10^{-3}$$

Figure 2.6. Noise equivalent temperature for a pushbroom at 200 km in two spectral regions.

The second line is the bandwidth calculation. The line time was found to be 1 ms, and the bandwidth is one divided by twice this value. It is a function of the integrated flux L_q and the change of the photance as a function of temperature, that is,

$$L_q = \int_{\lambda_1}^{\lambda_2} \frac{2c}{\lambda^4 \left(e^{\frac{c_2}{\lambda T}} - 1 \right)} d\lambda \qquad \frac{\partial L_q}{\partial T} = dL_q = \int_{\lambda_1}^{\lambda_2} \frac{2c \frac{c_2}{\lambda T} e^{\frac{c_2}{\lambda T}}}{\lambda^4 \left(e^{\frac{c_2}{\lambda T}} - 1 \right)^2 T} d\lambda \ . \qquad (2.14)$$

Then, the NETD is calculated as

$$NETD = \frac{1.35 \sqrt{L_q B}}{\lambda_{ave} dL_q} \ , \qquad (2.15)$$

and the average wavelength is calculated on the same line. The bottom half of the figure is a repeat of the same equations, but with the other wavelength limits. For the pushbroom, the NETD values are 0.033 and 0.00429, which are adequate values (lower than 0.1K), but may not remain adequate after all efficiencies are included.

2.5 OPTICS

We can start with the diffraction limit. Figure 2.7 shows the required resolution as a function of altitude and the required diameter as a function of altitude for both spectral regions. The first line of equations sets the domain of the altitude, R, from 100 to 1000 km in 1-km steps, the swath width, S, to 200 km, the ground spot to 10 m (0.01km), and the two wavelengths to 5 and 14 µm (0.0005 and 0.0014 cm). The next line calculates the two angular resolutions based on the earlier equations. The half-field angle is

$$\Theta = \arctan(S/2R) \ . \qquad (2.16)$$

Because the angle depends on the height, R, Mathcad requires the height to be shown. Mathcad also uses atan(x) for arctan(x). The slant range is

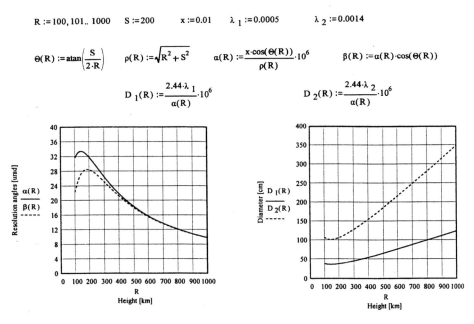

$$R := 100, 101 .. 1000 \qquad S := 200 \qquad x := 0.01 \qquad \lambda_1 := 0.0005 \qquad \lambda_2 := 0.0014$$

$$\Theta(R) := atan\left(\frac{S}{2 \cdot R}\right) \qquad \rho(R) := \sqrt{R^2 + S^2} \qquad \alpha(R) := \frac{x \cdot \cos(\Theta(R))}{\rho(R)} \cdot 10^6 \qquad \beta(R) := \alpha(R) \cdot \cos(\Theta(R))$$

$$D_1(R) := \frac{2.44 \cdot \lambda_1}{\alpha(R)} \cdot 10^6 \qquad\qquad D_2(R) := \frac{2.44 \cdot \lambda_2}{\alpha(R)} \cdot 10^6$$

Figure 2.7. Required resolution in μr and required aperture diameter in cm as a function of orbital altitude in km.

$$\rho = \sqrt{R^2 + S^2} \; . \tag{2.17}$$

The two resultant angular resolutions are

$$\alpha = \frac{x\cos(\Theta)}{\rho}, \qquad \beta = \alpha\cos(\Theta) \; . \tag{2.18}$$

The third line of equations is the calculation of the required aperture diameters for the two wavelengths, that is, for each, the diameter is given by

$$D_1 = \frac{2.44\lambda_1}{\beta}, \qquad\qquad D_2 = \frac{2.44\lambda_2}{\beta} \; , \tag{2.19}$$

using the wavelengths $\lambda_1 = 5$ μm and $\lambda_2 = 14$ μm.

It is clear that, based on my rule of thumb, a 1-m-diameter aperture is the largest that can be put into orbit;[1] any LWIR system will have to be at 200 km or below. The MWIR will have to be at 800 km or below. From the shape of the resolution curve and some astute guesses, 200 km seems to be the optimum altitude. It is also sufficiently high that the orbit will have a duration that is long enough (but orbital duration is outside the scope of this text).

As shown in Fig. 2.8, the aberrations can be calculated for a range of altitudes from 100 to 500 km, using the field-of-view equation, Eq. (1.1), the required-resolution equation, Eq. (1.2), and the blur equations, Eqs. (1.8) through (1.11). The first line sets the height domain, R, from 100 to 500 km, the spatial resolution, x, at 10 m (0.01 km), the optical speed, F, at 7, the optics diameter at 35 cm, and the maximum wavelength, λ, at 6 μm (0.0006 cm). Finally, the focal length is calculated from the optics diameter and optical speed, $f = FD_o$. The next line calculates the half field of view, Θ, as the arctan of $S/2R$; the required resolution, α, as x $\cos^2(\Theta)/R$; and the three aberrations, spherical, coma, and astigmatism. The third line calculates the linear size of the detector and the diffraction blur.

Figure 2.8(a) shows the required resolution, α, the spherical aberration, β_{SA}, coma, β_{CA}, astigmatism, β_{AA}, and the diffraction blur, β. The required resolution is the lowest, slanted line; next to it, and crossing at 200 km is the diffraction limit; spherical aberration is below this; but coma and astigmatism are too high. I adjusted the optics speed–a rather slow F/7-- until I reached this point.

Figure 2.8(b) shows the aberrations as a function of speed for an altitude of 200 km. The diffraction limit and the required resolution are the same. Coma and astigmatism are too high for reasonable speeds, but spherical aberration is satisfactory at a speed of F/6. We can use an uncorrected Schmidt system, i.e., a spherical mirror with the stop at the center of curvature and no corrector. Other systems that might be suitable are detailed in Appendix B.[2] We need a system that has approximately a 40-μrad resolution with a half-field angle of 26.5 degrees. Those systems that come close to satisfying these requirements are: at a resolution of 36 μrad, the Super Schmidt at F/1.25 on a curved image; at 20 μrad, the Schwarzschild, at 25 μrad, the Reflective Schmidt; and 49 μrad, the SEAL. The Super Schmidt has a refractive corrector, which excludes it from consideration for this application. Thus the three suitable design forms are the Schwarzschild, Reflective Schmidt, and the SEAL, all described in my earlier book, *Infrared Systems Design*.

The Reflective Schmidt and Schwarzschild require a curved field; the SEAL does not. All three have three mirrors. I think the choice here is the SEAL, but a firm decision would involve further analysis with Code V, Zemax, or another lens-design program. The SEAL requires a primary that is about five times the diameter of the aperture stop. The Schwarzschild might work nicely as an eccentric pupil system.

$R := 100, 101 .. 500$ $S := 200$ $x := 0.01$ $F := 7$ $D_0 := 35$ $f := F \cdot D_0$ $\lambda := 0.0006$

$\Theta(R) := atan\left(\dfrac{S}{2 \cdot R}\right)$ $\alpha(R) := \dfrac{x \cdot \cos(\Theta(R))^2}{R}$ $\beta_{SA} := \dfrac{1}{128 \cdot F^3}$ $\beta ca(R) := \dfrac{\Theta(R)}{16 \cdot F^2}$ $\beta aa(R) := \dfrac{\Theta(R)^2}{2 \cdot F}$

$a(R) := f \cdot \alpha(R)$ $\beta := \dfrac{2.44 \cdot \lambda}{D_0}$

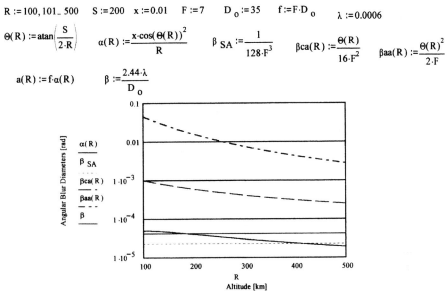

(a)

$R := 200$ $S := 200$ $F := 2, 2.1 .. 10$ $\lambda := 0.0006$ $D_0 := 35$ $\Theta := atan\left(\dfrac{S}{2 \cdot R}\right)$ $\alpha := \dfrac{x \cdot \cos(\Theta)^2}{R}$

$\beta_{SA}(F) := \dfrac{1}{128 \cdot F^3}$ $\beta_{CA}(F) := \dfrac{\Theta}{16 \cdot F^2}$ $\beta_{AA}(F) := \dfrac{\Theta^2}{2 \cdot F}$ $f(F) := F \cdot D_0$ $\beta := \dfrac{2.44 \cdot \lambda}{D_0}$

$a(F) := f(F) \cdot \alpha \cdot 10^4$

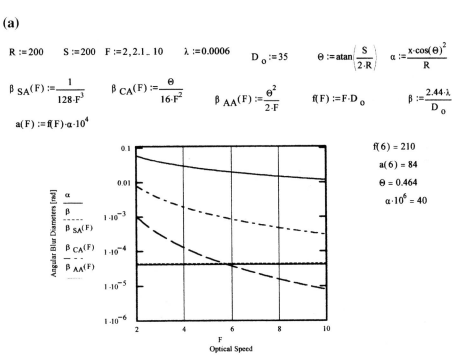

$f(6) = 210$

$a(6) = 84$

$\Theta = 0.464$

$\alpha \cdot 10^6 = 40$

(b)

Figure 2.8. a) Required resolution (rad) and blurs as a function of orbital altitude (km). b) Aberrations as a function of altitude and of optical speed.

2.6 THE STEP BACK

The system requirements of a large pushbroom field and a small angular pixel subtense resulted in rather extreme optical designs. This step back accounts for that, and recognizes that if these system requirements could be relaxed, the optical design constraints would be more flexible.

A Troika system, which uses three different optical telescopes to cover the field, might solve the problem. By applying Mathcad (as in Fig. 2.8), and changing the value of S, the lateral field, it is apparent that the Troika will not work. We must go to ten mirrors before things get simpler–simpler, yes; better, no.

2.7 EFFICIENCIES

Efficiencies to be addressed include scan efficiency, detector quantum efficiency, cold-shielding efficiency, optical efficiency, and atmospheric transmission.

2.7.1 Scan efficiency

For a pushbroom, the scan efficiency is 1.

2.7.2 Quantum efficiency

Mercury-cadmium-telluride detectors have a quantum efficiency of 0.8 and are good detectors for both spectral regions.

2.7.3 Cold-shield efficiency

The idealized equations assume that each detector element sees only the F/cone of the system. This is not the case with an array. Figure 2.9 shows the geometry, defines the symbols, and derives the cold-shielding efficiency for this one-dimensional case. Figure 2.10 shows the efficiency as a function of the position of the cold stop. The first line sets the aperture diameter D at 35 cm, the optical speed F at 4, calculates the focal length as FD, sets the distance L (as defined in Fig. 2.9) to 100, and sets the domain of s from 0 to 140 (in the same units). I assume that we do not want to complicate the system by a reimaging technique that would generate100% efficiency. We will take this value as 50%

2.7.4 Optics efficiency

Optics efficiency is the product of the reflectivity of the mirrors and transmissivity of the refractors. We assume a window for the detector housing, which, of course,

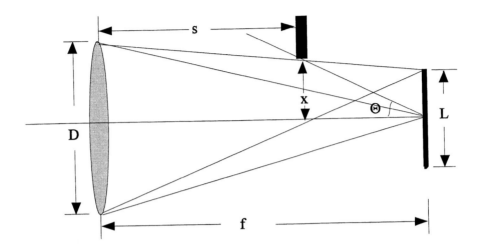

Figure 2.9. Geometry of the cold shield.

$$\text{num}(s) := \text{atan}\left(\frac{D}{2 \cdot f}\right) \qquad \text{den}(s) := \text{atan}\left[\frac{(D \cdot f - s \cdot (D - L))}{2 \cdot f \cdot (f - s)}\right] \qquad \eta_{cs}(s) := \left(\frac{\text{num}(s)}{\text{den}(s)}\right)$$

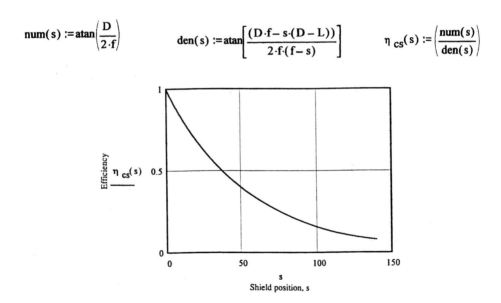

Figure 2.10. Cold shield efficiency as a function of shield position for a one-dimensional array.

must be at a temperature of about 80 K. Then each of the three mirrors of the SEAL or Schwarzschild or the two mirrors of the correctorless Schmidt is assumed to have a reflectivity of 95%. This value is conservative, representing a good overcoated aluminum or gold after reasonable aging. The detector window is assumed to be properly coated with a value of 90%. The optical efficiency then is either $0.95^3 \times 0.9 = 77\%$ or $0.95^2 \times 0.9 = 81\%$.

2.7.5 Atmospheric transmission

The atmosphere absorbs even in the good transmission regions. A proper and convenient way to calculate this is with PCModWin,[3] a commercial program that runs under Windows.™ When I used the slant path at the edge of the field and the desert model for the nature of the atmosphere, two interesting things arose: 1) the optimum spectral band is not 3 to 5 µm but 3.3 to 4.16 µm and, 2) the average value is 75%. So the efficiency needs to be taken as 75%, and the spectral limits of the sensitivity calculations must change.

2.7.6 Efficient summary

Figure 2.11 represents the new calculations. The first line sets the temperature at 300 K, the wavelengths at 3.3 and 4.16 µm, and introduces values for c and c_2. The next line is the bandwidth calculation, and then L_q and dL_q, the photance and temperature change of photance, are calculated with the integrals. The efficiency values are introduced in the next line: quantum efficiency, η of 0.8; scan efficiency, η_{sc} of 1; cold-shielding efficiency, η_{cs} of 0.5; optics transmission, τ_o of 0.77; and atmospheric transmission, τ_a of 0.75. The average wavelength and the NETD are calculated in the next line, with all the efficiency factors in the right places, which raises the NETD to a value that is higher than required, 0.191 K. Four rows of detectors will take care of this little problem, because the time-delay-and-integration (TDI) operation increases sensitivity by the square root of the number, or 0.095 K.

Table 2.1 compares the different pushbroom devices that have been discussed. Notice that the SEAL violates the 1-m-maximum-diameter rule of thumb. The Schmidt and Swarzschild are also larger than the aperture stop because they are not at the aperture stop and the field is not zero. Although the systems are similar, I believe the correctorless Schmidt is the best choice. It has fewer elements, lower NETD, and is simpler to manufacture and align.

2.8 THE WHISKBROOM

The pushbroom has the wonderful advantage of no moving parts, but the almost pathological optical requirements of high resolution and large field of view led to large designs. Perhaps a little motion will alleviate this problem.

Unlike the pushbroom, which has an array of detectors that covers the width of the full field of view and relies solely on the motion of the vehicle for scanning, the whiskbroom uses a mirror to sweep a smaller field of view from side to side to cover the full field.

$$T := 300 \qquad \lambda_1 := 0.00033 \qquad \lambda_2 := 0.000416 \quad c_2 := 1.4388 \qquad c := 2.9975 \cdot 10^{10}$$

$$B := \frac{1}{2 \cdot 0.001}$$

$$L_q := \int_{\lambda_1}^{\lambda_2} \frac{2 \cdot c}{\lambda^4 \cdot \left(e^{\frac{c_2}{\lambda \cdot T}} - 1 \right)} \, d\lambda$$

$$dL_q := \int_{\lambda_1}^{\lambda_2} \frac{2 \cdot c \cdot \frac{c_2}{\lambda \cdot T} \cdot e^{\frac{c_2}{\lambda \cdot T}}}{\lambda^4 \cdot \left(e^{\frac{c_2}{\lambda \cdot T}} - 1 \right)^2 \cdot T} \, d\lambda$$

$$\eta := 0.8 \qquad \eta_{sc} := 1 \qquad \eta_{cs} := .5 \qquad \tau_o := 0.77 \qquad \tau_a := 0.75$$

$$\lambda_{ave} := \frac{\lambda_1 + \lambda_2}{2} \qquad\qquad NETD := \frac{1.35 \sqrt{\dfrac{L_q \cdot B}{\eta_{sc} \cdot \eta_{cs} \cdot \eta}}}{\tau_a \cdot \tau_o \cdot \lambda_{ave} \cdot dL_q}$$

$$NETD = 0.191$$

Figure 2.11. **Noise-equivalent-temperature differences for a SEAL or Schwarzchild pushbroom at 200 km in the revised spectral region.**

Table 2.1. Pushbroom systems.

Property	Units	Correctorless Schmidt	SEAL	Schwarzschild
Spectrum	μm	3.33-4.16	3.33-4.16	3.33-4.16
Altitude	km	200	200	200
Field	deg	53	53	53
Resolution	μrad	40	40	40
Aperture Diameter	cm	35	35	35
Mirror Diameter	cm	55	175	55
Optical Speed	F#	6	4	4
Focal Length	cm	210	150	150
Overall Length	cm	220	50	100
Detector Size	μm	84	60	60
Detector Number	--	20300x4	20300x4	20300x4
Array Size	cm	171x0.034	122x0.034	122x0.034
Optics Efficiency	%	86	77	82
Cold Shield Efficiency	%	50	50	50
Scan Efficiency	%	100	100	100
NETD	K	0.081	0.095	0.095

2.8.1 GEOMETRY

The whiskbroom geometry is dictated by the kind of array that is assumed. For example, a reasonable assumption is four rows of 512 mercury-cadmium-telluride detectors arranged parallel to the direction of travel, in the across-track direction. From this, the resolution and field of view are not difficult to estimate. The field is certainly too small for the cosine effect to be important. Then the 0.01-km spot subtends an angle of 0.01/200 = 50 μrad, and 256 spots yield a half-field angle of 0.0128 rad or 0.73 degrees. This is far from the optics requirements of the pushbroom. Figure 2.12 illustrates the geometry.

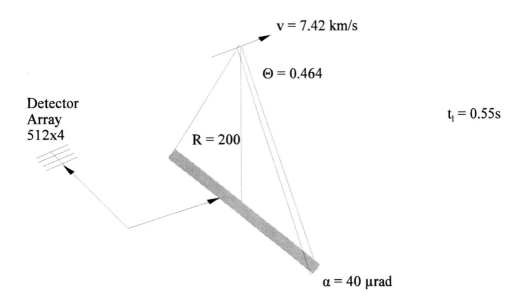

Figure 2.12. Geometry and parameters of the whiskbroom.

2.8.2 Dynamics

The line time for the whiskbroom is longer than that of the pushbroom by a factor of 512--that is, 0.55 s. The dwell time and bandwidth are calculated in Figure 2.13. The first line inputs the temperature, T=300K; the short and long wavelength limits, $\lambda_1 = 0.000333$ cm and $\lambda_2 = 0.000416$ cm; $c_2 = 1.4388$ cmK and c=2.9975x1010 cm/s. The second line inputs the ground spot, x=0.01 km; the altitude domain, R=100, 101...500km; the orbital velocity, v = 7.42; the swath width, 200 km; the number of across-track detectors in the array, m_a=512; and, finally, the in-track number of detectors, m_i = 4. In the next line, the half-field, Θ, is calculated as Θ=atan(S/2R); the most difficult resolution angle, α, as $x\cos^2(\Theta)/R$; and the number of resolution elements in a full field, N = Θ/α. The line after includes calculation of the line time, $t_l = \alpha R m_a/v$; the dwell time, $t_d = t_l/N$; and the bandwidth, B=1/2$t_d m_i$. The two integral calculations, Lq and dLq, are to be entered into the NETD equation with the average wavelength. The resolution, α, half field, Θ, and line time, t_l, for an altitude of 200 km are displayed as 40 µrad, 0.464, and 0.552 s, respectively. The two curves display the NETD and dwell time as a function of altitude.

$$T := 300 \qquad \lambda_1 := 0.000333 \quad \lambda_2 := 0.000416 \quad c_2 := 1.4388 \qquad c := 2.9975 \cdot 10^{10}$$

$$x := 0.01 \qquad R := 100, 101 .. 500 \qquad v := 7.42 \qquad S := 200 \qquad m_a := 512 \quad m_i := 4$$

$$\Theta(R) := \operatorname{atan}\left(\frac{S}{2 \cdot R}\right) \qquad \alpha(R) := \frac{x}{R} \cdot \left(\cos(\Theta(R))^2\right) \qquad\qquad N(R) := \frac{\Theta(R)}{\alpha(R)}$$

$$t_l(R) := \frac{\alpha(R) \cdot R \cdot m_a}{v} \qquad t_d(R) := \frac{t_l(R)}{N(R)} \qquad B(R) := \frac{1}{2 \cdot t_d(R) \cdot m_i}$$

$$L_q := \int_{\lambda_1}^{\lambda_2} \frac{2 \cdot c}{\lambda^4 \cdot \left(e^{\left(\frac{c_2}{\lambda \cdot T}\right)} - 1\right)} d\lambda \qquad\qquad dL_q := \int_{\lambda_1}^{\lambda_2} \frac{2 \cdot c \cdot \frac{c_2}{\lambda \cdot T} \cdot e^{\frac{c_2}{\lambda \cdot T}}}{\lambda^4 \cdot \left(e^{\frac{c_2}{\lambda \cdot T}} - 1\right)^2 \cdot T} d\lambda$$

$$\lambda_{ave} := \frac{\lambda_1 + \lambda_2}{2} \qquad NETD(R) := \frac{1.35 \cdot \sqrt{L_q \cdot B(R)}}{\lambda_{ave} \cdot dL_q} \qquad \alpha(200) = 4 \cdot 10^{-5} \qquad t_l(200) = 0.552$$

$$\Theta(200) = 0.464$$

$$B(200) = 2.625 \cdot 10^3$$

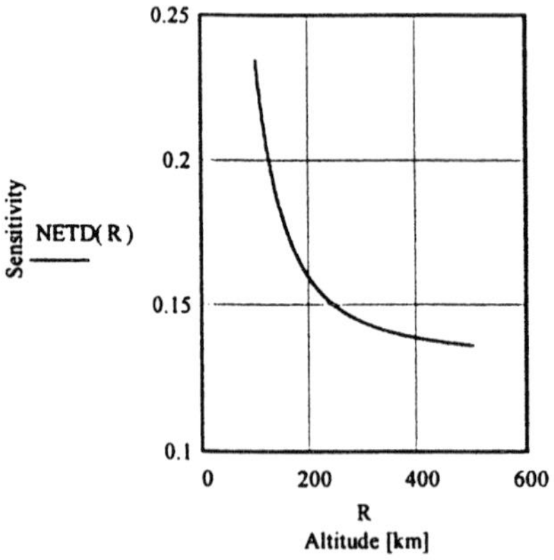

$$NETD(200) = 0.159$$

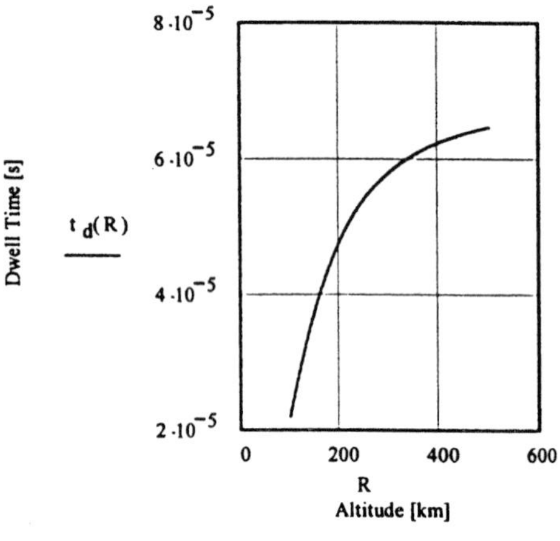

Figure 2.13. Whiskbroom sensitivity and geometry with a 512x4 array.

2.8.3 Sensitivity

Figure 2.13 provides the sensitivity calculations including the calculation of the line time, the dwell time, the bandwidth, and the corrected spectrum. Because we cannot do all of the efficiencies yet, we do none. As Fig. 2.13 shows, the dwell time is longer than the response time of the detectors. (Whew!) It also shows that the NETD is too high (even without efficiency factors). We are sufficiently close to evaluate with efficiency factors, and are ready to improve efficiency by adding detectors.

2.8.4. Optics

Figure 2.14 shows the usual optics calculations. The altitude R=200 km; the swath width S=200 km; the resolution spot x=0.01 km; there are 512 across-track detectors, m_a; and the optical speed is from F=3 to F=10. The half field is calculated in the usual way, Θ=atan($S/2R$); The required resolution as α=xcos2Θ/r and the half-field angle for the array, θ=$m_a\alpha$. It is the half-field angle of the array that sets the requirements for a whiskbroom. The graph shows, from the top down at the y axis, spherical aberration, coma, astigmatism, and the requirement. The aberrations are small enough for a speed of about F/5.5--which appears reasonable--a simple spherical mirror is the optical system. But we must consider the obscuration of the detector array and all its electronics and support. We must get the detector array out of the path of the incoming light, and classical designs do this: spherical Herschellian, Newtonian, Pfundian, and eccentric-pupil sphere. The Herschellian works at a field angle; the Newtonian suffers from obscuration; the Pfundian requires a large folding mirror; and the eccentric pupil requires twice the curvature. (Tradeoffs, tradeoffs.)

Figure 2.15 shows the Herschellian, with an off-axis angle of about 7 degrees. This must be added to the half-field angle for a recalculation of the aberrations. The results are disastrous. Figure 2.16 illustrates the Newtonian, which indicates a small obscuration (about 1%). Figure 2.17 shows the Pfundian and its large folding flat. Figure 2.18 shows the eccentric-pupil system and the mirror from which it is obtained. Each of these mirrors is a parabola. My choice would be the eccentric-pupil system. It is lighter, has a higher optical efficiency, no obscuration, and the F/4 requirement means fabrication of an F/2 system (which is readily possible).

The scan mirror must be about 1.5 times the diameter of the aperture diameter (52.5 cm) and must rotate at about 1.7 cy/s. The edge rotates at about 1 kph or 0.6 mph, which is acceptable.

$R := 200$ $S := 200$ $x := 0.01$ $m_a := 512$ $F := 3, 3.1 .. 10$

$$\Theta := \text{atan}\left(\frac{S}{2 \cdot R}\right) \qquad \alpha := \frac{x}{R} \cdot \cos(\Theta)^2 \qquad \theta := m_a \cdot \alpha$$

$$\beta_{SA}(F) := \frac{1}{128 \cdot F^3} \qquad \beta_{CA}(F) := \frac{\theta}{16 \cdot F^2} \qquad \beta_{AA}(F) := \frac{\theta^2}{2 \cdot F}$$

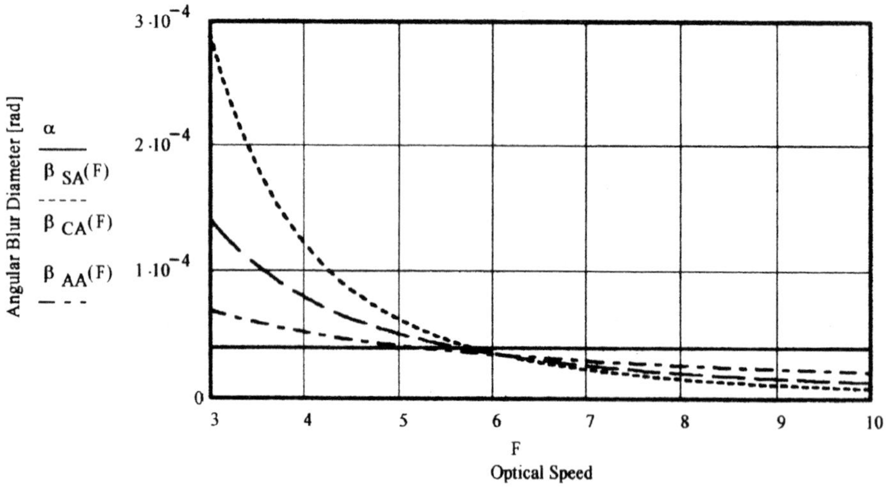

Figure 2.14. Aberrations of the whiskbroom system.

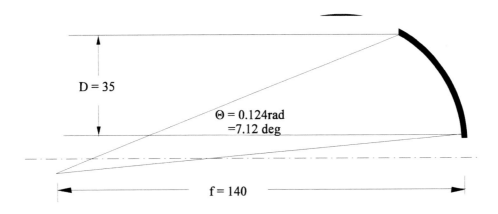

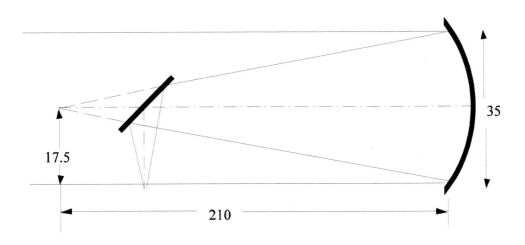

Figure 2.15. The herschellian telescope.

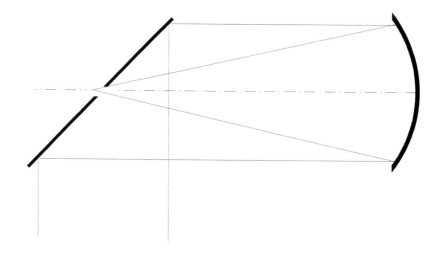

Figure 2.16. The newtonian system with 6.25% obscuration.

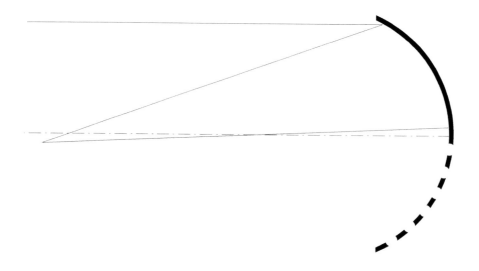

Figure 2.17. The pfundian system with its large folding mirror.

2.8.5 Efficiencies

All of the efficiencies differ from the pushbroom, except that the atmospheric transmission and detector quantum efficiency remain at 75% and 80%, respectively.

The cold-shielding efficiency for a single row of 512 detectors is up to 75%.

The optics efficiency is the product of the reflectance of a single mirror and the scan mirror, and the transmittance of the detector-housing window. In other words, 95%, 95%, and 90%--for a total of 81%.

The scan efficiency is drastically different. If the mirror rotates, it travels through a full circle, but only 0.928 radians is useful, yielding 14.8%.

The results of combining these efficiencies, and using the correct spectral region, are shown in Fig. 2.19. The input data are at the top: $T=300K$, $\lambda_1 = 0.000033$ μm, $\lambda_2 = 0.000416$ μm, $c_2 = 1.4388$, $c=2.9975 \times 10 10$, and $B= 2626$. The integrals for L_q and dL_q are evaluated and the efficiencies are input: $\eta=0.8$, $\eta_{sc} = 0.148$, $\eta_{cs} = .75$, $\tau_o = 0.81$ and $\tau_a = 0.75$. The NETD is calculated as before with the average wavelength and the factors. It is 0.883K, a factor of almost 9 too high. Thus we need more detectors--a TDI set that numbers 81, the square of the ratio. That is a very long TDI set; it might work, but we need to STEP BACK.

2.8.6 The step back

The whiskbroom, although large, is a nice, simple system. The scan mirror rotates at 1.7 cy/s. Although the scan efficiency of the design is problematic, we can improve it by oscillating the mirror. It is not easy to oscillate a large mirror, but it is possible. The mirror may be considered circular (or oval) with a thickness of about one fifth the diameter and lightweighted 50%. The volume is π x 52.5^2 x $52.5/10 = 45,460$ cm^3. If the mirror is made of beryllium, with a specific gravity of 1.85, then the mirror weighs 84.1 kg, or 185 lbs. Although that is more than I weigh, with the proper mechanism we can indeed oscillate the beryllium scan mirror. If the scan efficiency is increased from 14.8% to an estimated 90%, the detector TDI stages can be reduced from 81 to 14–a reasonable value.

2.8.7 The whiskbroom summary

The properties of the whiskbroom scanner are summarized in Table 2.2. Even though the whiskbroom moves, it appears to be a better solution than the others.

$T := 300$ $\lambda_1 := 0.00033$ $\lambda_2 := 0.000416$ $c_2 := 1.4388$ $c := 2.997510$

$B := 2625$

$$L_q := \int_{\lambda_1}^{\lambda_2} \frac{2 \cdot c}{\lambda^4 \cdot \left(e^{\frac{c_2}{\lambda \cdot T}} - 1 \right)} \, d\lambda$$

$$dL_q := \int_{\lambda_1}^{\lambda_2} \frac{2 \cdot c \cdot \frac{c_2}{\lambda \cdot T} \cdot e^{\frac{c_2}{\lambda \cdot T}}}{\lambda^4 \cdot \left(e^{\frac{c_2}{\lambda \cdot T}} - 1 \right)^2 \cdot T} \, d\lambda$$

$\eta := 0.8$ $\eta_{sc} := .148$ $\eta_{cs} := .75$ $\tau_o := 0.81$ $\tau_a := 0.75$

$$\lambda_{ave} := \frac{\lambda_1 + \lambda_2}{2}$$

$$NETD := \frac{1.35 \sqrt{\dfrac{L_q \cdot B}{\sqrt{\eta_{sc} \cdot \eta_{cs} \cdot \eta}}}}{\tau_a \cdot \tau_o \cdot \lambda_{ave} \cdot dL_q}$$

$NETD = 0.883$

Figure 2.19. The sensitivity of the whiskbroom system with efficiencies.

Table 2.2. Properties of the whiskbroom scanner.		
Property	Units	Eccentric Pupil
Spectrum	3.33	4.16
Altitude	km	200
Field	deg	0.146
Resolution	μrad	40
Aperture Diameter	cm	35
Mirror Diameter	cm	35
Scan Mirror Diameter	cm	52.5
NETD	K	0.1
Optical Speed	F#	6
Focal Length	cm	210
Overall Length	cm	210
Detector Size	μm	105
Detector Number	--	512x14
Array Size	cm	5.37
Optical Efficiency	%	81
Cold-Shield Efficiency	%	75

2.9 THE STEP STARER

The final system to be considered is a step-staring one. We will put a square array of detectors at the focal plane, and aim it at one part of the field, and then the next, and then the next, and so on. This will require a gimbaled system, with either moving optics or a moving pointing mirror. The pointing accuracy should be about one pixel (about 40 μrad), because a few seams in the overall image are acceptable in the final image.

2.9.1 Geometry and dynamics

The required resolution must be calculated based on the corner pixels. The geometry in Fig. 2.20 shows that the full half field is 0.616 rad and the resolution angle is 33.3 μrad. The field is 36,.997 pixels. If an array with 512x512 elements is used, then 72x72 individual fields must be used. The available frame time is $200/7.42 = 27$ s. This leaves 5.78 ms for each field, including 3 s, or about 10%, for all switches from field to field. The field and subfields are summarized in Fig. 2.21.

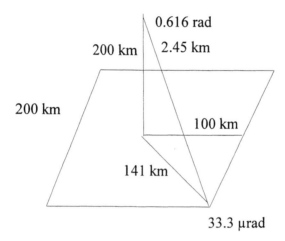

Figure 2.20. The geometry of the step starer.

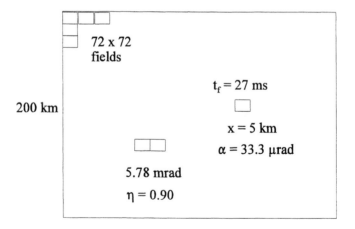

Figure 2.21. Schematic of the step staring system.

2.9.2 Sensitivity

The sensitivity calculations are shown in Fig. 2.22. Because this array is larger than the whiskbroom array, the NETD is more than satisfactory at 0.028K.

$$T := 300 \qquad \lambda_1 := 0.00033 \qquad \lambda_2 := 0.000416 \qquad c_2 := 1.4388 \qquad c := 2.9975 \cdot 10^{10}$$

$$t_d := 0.027 \qquad B := \frac{1}{2 \cdot t_d}$$

$$L_q := \int_{\lambda_1}^{\lambda_2} \frac{2 \cdot c}{\lambda^4 \cdot \left(e^{\frac{c_2}{\lambda \cdot T}} - 1 \right)} \, d\lambda$$

$$dL_q := \int_{\lambda_1}^{\lambda_2} \frac{2 \cdot c \cdot \frac{c_2}{\lambda \cdot T} \cdot e^{\frac{c_2}{\lambda \cdot T}}}{\lambda^4 \cdot \left(e^{\frac{c_2}{\lambda \cdot T}} - 1 \right)^2 \cdot T} \, d\lambda$$

$$\eta := 0.8 \qquad \eta_{sc} := .9 \qquad \eta_{cs} := .75 \qquad \tau_o := 0.95 \cdot 0.9 \qquad \tau_a := 0.75$$

$$\lambda_{ave} := \frac{\lambda_1 + \lambda_2}{2} \qquad\qquad NETD := \frac{1.35 \sqrt{\dfrac{L_q \cdot B}{\eta_{sc} \cdot \eta_{cs} \cdot \eta}}}{\tau_a \cdot \tau_o \cdot \lambda_{ave} \cdot dL_q}$$

$$NETD = 0.028$$

Figure 2.22. Sensitivity of the step starer.

2.9.3 Optics

The optics requirements are more stringent than those of the whiskbroom and the calculations are shown in Fig. 2.23. The same format as that used in the earlier optics calculations is used. At an altitude of 200 km, the required resolution, the diffraction-limited resolution, and the spherical aberration blur are the same. Coma and astigmatism are unacceptably high.

$R := 100, 101 .. 500$ $S := 200$ $x := 0.01$ $F := 6$ $D_0 := 35$ $f := F \cdot D_0$ $\lambda := 0.0006$

$\Theta(R) := atan\left(\dfrac{S}{2 \cdot R}\right)$ $\alpha(R) := \dfrac{x \cdot \cos(\Theta(R))^2}{R}$ $\beta_{SA} := \dfrac{1}{128 \cdot F^3}$ $\beta ca(R) := \dfrac{\Theta(R)}{16 \cdot F^2}$ $\beta aa(R) := \dfrac{\Theta(R)^2}{2 \cdot F}$

$a(R) := f \cdot \alpha(R)$ $\beta := \dfrac{2.44 \cdot \lambda}{D_0}$

$a(200) = 8.4 \cdot 10^{-3}$

$R_0 := 2 \cdot f$

$R_0 = 420$

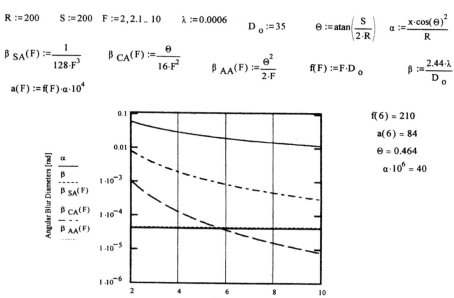

(a)

$R := 200$ $S := 200$ $F := 2, 2.1 .. 10$ $\lambda := 0.0006$ $D_0 := 35$ $\Theta := atan\left(\dfrac{S}{2 \cdot R}\right)$ $\alpha := \dfrac{x \cdot \cos(\Theta)^2}{R}$

$\beta_{SA}(F) := \dfrac{1}{128 \cdot F^3}$ $\beta_{CA}(F) := \dfrac{\Theta}{16 \cdot F^2}$ $\beta_{AA}(F) := \dfrac{\Theta^2}{2 \cdot F}$ $f(F) := F \cdot D_0$ $\beta := \dfrac{2.44 \cdot \lambda}{D_0}$

$a(F) := f(F) \cdot \alpha \cdot 10^4$

$f(6) = 210$

$a(6) = 84$

$\Theta = 0.464$

$\alpha \cdot 10^6 = 40$

(b)

Figure 2.23.   Required resolution (rad) and blurs as a function of orbital altitude (km) and (b) aberrations as a function of altitude and of optical speed.

However, an F/6 correctorless Schmidt does the job! The mirror radius of curvature is 420 cm–a rather small curvature. It may be necessary to use a lens near the focus to flatten the field.

I envision the optical system mounted on a gimbal that points successively to each of the subfields. Therefore the optical efficiency is that of a mirror and a detector window and, perhaps, a field flattener.

2.9.4 The step back

The driving requirement here is the imaging of a 10-m spot in the corner of the field, based on the geometry described. This led to a resolution requirement of 33.3 μrad and a slower optical system. But another possibility is for us to consider a step starer that "scans" lines. Perhaps a way to think of it is that it is a "jerky" whiskbroom.

The line time is 0.69 s. At the edge of the linear field, the 10-m spot subtends 40 μrad. There are 23,000 pixels and therefore a required 45 subfields in the line. Each subfield will take 15 ms. This is just fine, even with 10 % of the time between successive subfields. If each field takes 10 ms, then there are 5 ms to scan the gimbaled optics a total of 21 mrad to the next field, a rate of 4 rad/s, but a distance of only 21 mrad.

2.10 DESIGN SUMMARY

Table 2.3 summarizes the step starer with the best pushbroom and best whiskbroom designs. So what is best among the best pushbroom, best whisk broom and best step-starer? The NETDs are comparable. The pushbroom has a larger mirror. The jerky system and the step starer have the most detector elements, but they are available arrays. The pushbroom would be constructed of segments of something like 1024x4 arrays. The overall (unfolded) length of the jerky and stepper is larger than that of the brooms. The pushbroom uses a correctorless Schmidt; the stepper and jerky use a simple sphere. The pushbroom has no moving parts. The jerky and stepper move mildly with a gimbal. The copout is to leave the choice to the mechanical engineers--the next generation. I will.

2.11 VARIATIONS

The Peacekeeper shell game was used to illustrate the design of a very challenging strip mapper. It was described to push the limits and illustrate design principles, to go all out in all the ways that might get there. Other less-challenging strip-mapper designs include some of the orbiting systems that have been used for remote sensing: Landsat, Thematic Mapper, and Satellite Pour l'Observation de la Terre (SPOT), to name just three. The military used these mappers in low-flying aircraft long before the advent of the so-called FLIRs, and continues to use them today.

Table 2.3 Summary of summaries: The choice system.

Property	Units	Pushbroom	Whiskbroom	Jerky System	Step-Starer
Spectrum	μm	3.3-4.16	3.3-4.16	3.3-4.16	3.3-4.16
Altitude	km	200	200	200	200
Field	deg	53	1.17	1.17	0.976
Resolution	μrad	40	40	40	33.3
Aperture Diameter	cm	35	35	35	35
Mirror Diameter	cm	175	35	35	35
Scan Mirror	cm	--	50	--	--
Optical Speed		4	4	6	6
Focal Length	cm	140	140	210	210
Overall Length	cm	150	150	220	220
Detector Size	μm	56	105	105	70
Detector Number		20300x4	512x14	512x512	512x512
Array Size	cm	114x0.02	5.38x0.147	5.38x5.38	3.6x3.6
Optics Efficiency	%	77	81	86	86
Cold-Shield Efficiency	%	50	75	75	75
Scan Efficiency	%	100	90	90	90
NETD	K	0.095	0.098	0.028	0.028

2.11.1 Landsat, thematic mapper, and multispectral scanner

The properties of the Thematic Mapper and its predecessor, the Multispectral Scanner for Landsat satellites, are summarized in Table 2.4. The Thematic Mapper flies in a sun-sychronous orbit (at 9:30 a.m.) at an altitude of 705 km, has a 0.41-m-diameter Ritchey-Chretien telescope with a speed of F/5.6, a scan frequency of 7 Hz, a GSD of 120 m, and a ground swath of 185 km. Thus, the angular resolution is 0.17 mrad. The major difference between this and the MX mapper is a factor of 6 in the number of elements and different spectral bands (see Table 2.4). The Multispectral Scanner has lower performance than the TM, and the "evenness" of the bands indicates the earlier state of the art before investigators had honed the bands for certain interpretations. It also crosses the equator at 9:30 a.m. in a sun-synchronous orbit at 919 km.

Band	Spectrum	GSD	NERD[%] NETD[K]	Band	Spectrum	GSD	NERD[%] NETD[K]
Table 2.4 Characteristics of two remote sensors.							
1	0.45-0.52	30	0.65				
2	0.52-0.60	30	0.57				
3	0.63-0.69	30	0.57				
4	0.76-0.90	30	0.33	4	0.5-0.6	76	0.57
5	1.55-1.75	30	1.68	5	0.6-0.7	76	0.57
6	10.4-12.5	120	0.5K	6	0.7-0.8	76	0.65
7	2.08-2.35	30	2.0	7	0.8-1.1	76	0.7
8				8	10.4-12.6	234	1.4K

2.11.2 The SPOT

The satellite pour l'observation de la terre (SPOT) is a French satellite that was first launched in 1986. It flies at a mean altitude of 832 km in a sun-synchronous circular orbit and has a swath width of 117 km. The optics consist of a Bouwers-Maksutov telescope and several arrays of colored and black and white detectors.

Note the differences between these remote sensors and our rather difficult problem. The NETDs here are 5 and 14 times higher (easier). The ground spot is 7.6 and 12 times that of the MX device. The ground swath is smaller by about 10%, and the spectral region is LWIR. All of these factors make the design of the SPOT easier.

2.11.3 The spy satellite

Another interesting calculation that can be made based on the material presented here, is whether the spy satellite that was described by Tom Clancy can actually read license plate numbers. To assess this, we assume that the Hubble telescope–the largest telescope in orbit-- is flown at an altitude of 100 km and turned to look down at my backyard. If the array consists of silicon elements that operate in the visible, then the longwave limit may be assumed to be 1 μm. Thus the resolution spot will be

$$\alpha = \frac{2.44\lambda}{D} = \frac{2.44 x 10^{-6}}{3} = 0.813 [\mu rad] \ . \qquad (2.20)$$

The resolution on the ground is then 0.08 m or 8 cm, which is about the height of one of my plate numbers, and the width is much smaller, about 1 cm. According to the Johnson criteria, one must have almost five pixels across the number to read it, which requires a resolution of 2 mm--a factor of 40 from that calculated above. Even limiting the spectral band to a maximum of 0.5 μm doesn't yield the desired results, and no amount of image processing would easily bridge the gap. I am willing to believe that the Clancy spy satellite can see me or my guests sun bathing in my backyard (in any state of undress), but they cannot read my license plate--especially because it is mounted vertically!

2.12 REFERENCES

[1]The Hubble telescope is larger, but such investments in time and money will probably never be repeated.

[2]L. Jones, Chapter 18, in M. Bass, E. Van Stryland, D. Williams, W. Wolfe, eds. *The Handbook of Optics Vol. II*, McGraw-Hill (1995).

[3]Ontar Corporation, North Andover, MA 01845-2000.

CHAPTER 3
THE SPACE CAMPOUT

The space campout is a thinly veiled representation of the detection of ballistic missiles in their midcourse phase. It has a certain whimsy that I like, and it sounds nicer.

3.1 PROBLEM STATEMENT

Two astronauts were camping out in the Shuttle. Happy hour was approaching, but their block of ice had disappeared. This was a 1-m^3 block from which they chipped the ice that would cool their drinks. The astronauts hoped to find it quickly--within the next 20 minutes. So they pulled out their pads and pencils to design an infrared device to detect and locate the precious chunk of hard water. If they succeed, it will surely be the fastest infrared in space!

3.2 THE GEOMETRY

We will assume that the Shuttle altitude, h, is 1000 km, that is, 1 Mm. It is in circular orbit around the Earth. The block is probably at the same altitude, but that cannot be assumed. We can assume that the block does not go higher. We will also assume that it separated from the Shuttle at a differential velocity. Figure 3.1 illustrates the geometry. The radius of the earth, R_E, is 6378 km, the orbital altitude, h, is 1000 km, and the angle, θ, is determined by

$$\theta = \arctan\left(\frac{a}{\sqrt{(R_E+h)-R_E^2}}\right) = \arctan\left(\frac{a}{R}\frac{1}{\sqrt{\frac{2R_E}{h}+1}}\right). \tag{3.1}$$

The range, R, is found from

$$R = \sqrt{2hR_E+h^2+a^2} = h\sqrt{\frac{2R_E}{h}+\left(\frac{a}{h}\right)^2+1}. \tag{3.2}$$

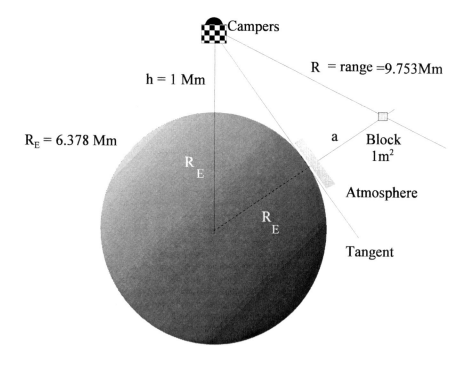

Figure 3.1. Geometry of the space campout. R_E is the radius of the Earth; R is the range; h is the orbital altitude, and a is the altitude of the block.

Figure 3.2 shows how the field of view and the range change with the assumed altitude of the block of ice. The mean radius of the Earth is 6378 km, as shown in the first line. As a function of the altitude, a, of the block of ice above the tangent point, from 100 to 1000 km, the field angle is given by Eq. (3.1) and the range R is given by Eq. (3.2). They are the next line in the Mathcad layout. For an assumed altitude, a, of 1000 km, the field is 0.263 rad and the range is 9753 km. But this is the field from the surface of the Earth to the altitude and the range to the perpendicular to the tangent. The field can be smaller, and the range should be greater. For first iteration purposes, assume that the range is 10 Mm, a little more than calculated, and the field is 0.1 rad, less than half that just calculated. The block of ice, measuring 1 m on a side, subtends 0.1 μrad at 10 Mm. If we attempted to match the subtense of the block with an image of the detector, there would be approximately one million detectors in a row. Fortunately, this is not necessary. This is a detection problem, and only detection must be performed. The resolution angle can be much larger. In fact, if the block must be found with a

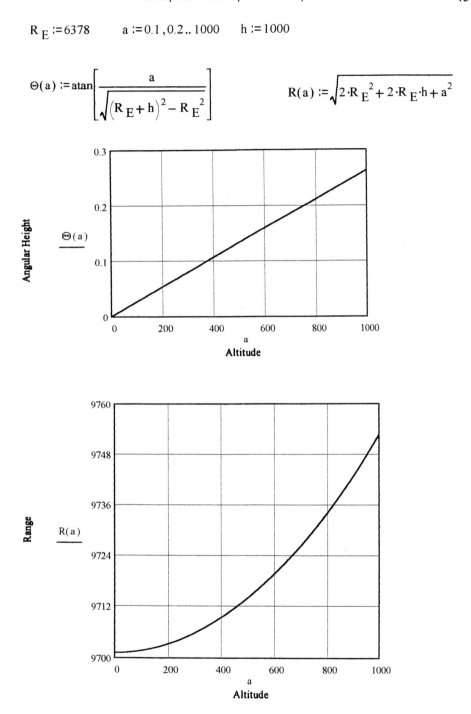

$R_E := 6378 \qquad a := 0.1, 0.2 .. 1000 \qquad h := 1000$

$$\Theta(a) := \operatorname{atan}\left[\frac{a}{\sqrt{\left(R_E + h\right)^2 - R_E^2}}\right] \qquad\qquad R(a) := \sqrt{2 \cdot R_E^2 + 2 \cdot R_E \cdot h + a^2}$$

Figure 3.2. Field of view and range as a function of the altitude of the block.

positional uncertainty no larger than 1 km, then the angular resolution must be about 0.1 mrad. The full field will be 1000 times this or 0.1 rad, about 6 degrees. We can then scan the vertical field all around the earth in what has been called a coolie-hat scan, shown schematically in Fig. 3.3.

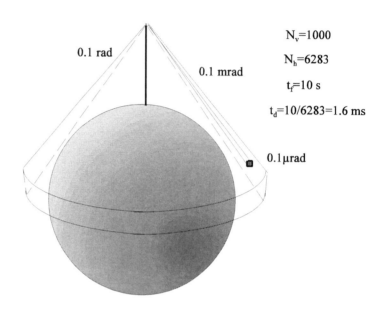

Figure 3.3. The coolie-hat scan and its parameters.

3.3 DYNAMICS

The scan must be done in less than 20 minutes. Thus, there must be 2π radians scanned by 1-mrad pixels in less than 20 minutes. The dwell time would be $(2 \times 3.14)/(0.001) \times (20 \times 60) = 7.54$ seconds. But this is far too slow. There would be only one detection of the block during that 20-minute period. We could probably scan at the rate of one per second to detect it early, get a few looks, and predict its trajectory. But one per second is faster than necessary.

Here we find some tradeoff space. As a first cut, assume that each scan takes 10 s. Then the dwell time is reduced by the ratio of 20 min (1200 s) to 10 s, or 120. The dwell time is 0.063 s; the bandwidth is therefore 7.95 or 8 Hz.

3.4 SENSITIVITY

We shall see that the sensitivity in this situation depends on the detector flux from the background. We can start, however, with the idealized equation.

3.4.1 The idealized equation

We must calculate the signal-to-noise ratio (SNR), for this problem. We can assume that the block of ice is at its normal freezing temperature (273 K) and has an emissivity[1] of 1. We know most of the remaining information needed for the SNR equation:

$$SNR = \varepsilon \tau_a \tau_o \sqrt{\eta} \frac{A_s A_o}{R^2 \sqrt{A_d B}} \frac{\int L_q d\lambda}{\sqrt{\int E_q d\lambda}} , \qquad (3.3)$$

where ε is the emissivity of the block, τ_a is the atmospheric transmission, τ_o is the optics transmission, A_s is the source area, A_o is the optical aperture, R is the range, B is the bandwidth, L_q is the photon radiance of the source, and E_q is the photon incidence on the detector. We choose to keep the line of sight above the atmosphere so that the transmittance, τ_a, is one. The optics transmission must be evaluated after we have an idea of what the optics are, but a reasonable beginning value is 0.5. The source area is 1 m^2. We can choose the optics area based on the diffraction limit at the selected wavelength. The 8- to 15-μm band is a good one to try because the 273-K block of ice peaks at 10.6 μm; but remember that no transmission loss occurs along the path we have chosen. Figure 3.4 shows the diffraction-limited diameter as a function of wavelength; a diameter of between 30 and 40 cm is sufficient, as seen in Eq. (3.4). The factor of 10,000 converts the wavelength from centimeters to micrometers. The area of the optics can be taken as $(\pi/4) \, 0.34^2 \approx 0.1$ m^2.

$$D = \frac{2.44\lambda}{\alpha} = \frac{2.44 x 14}{0.0001} = 341600 \, [\mu m] = 0.34 [m] . \qquad (3.4)$$

The detector dimensions can be found by taking the optics diameter of 34 cm, and an assumed optical speed of 3 to obtain a focal length of 102 cm with the required

$\lambda := 0.0003, 0.00031 .. 0.0015 \qquad \alpha := 0.0001 \qquad F := 3$

$$D(\lambda) := \frac{2.44 \cdot \lambda}{\alpha} \qquad a(\lambda) := F \cdot D(\lambda) \cdot \alpha \cdot 10000$$

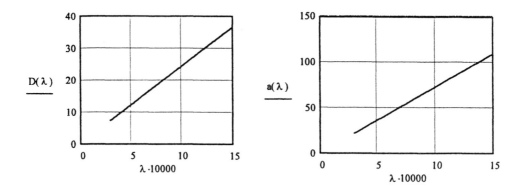

Figure 3.4. Required aperture diameter (in m) and detector size (in μm) for an F/3 system and 0.1-mrad resolution as a function of wavelength.

0.1-mrad resolution; it is 0.0102 cm or 102 μm. This value is well within the technology, and it provides an array length of 10 cm, also reasonable, as shown in Fig. 3.5. The range was taken as 10 Mm; the bandwidth as 8 Hz. The radiance of the source in this spectral band is evaluated in Fig. 3.5 as 2.685×10^{17}, and the SNR is

$$SNR = 1 x 1 x 0.5 x 0.9 \frac{1m^2 x 0.09 m^2}{(9.73 x 10^6 m)^2} \frac{1}{102 \mu m \sqrt{157 s^{-1}}} \frac{2.685 x 10^{17} (s cm^2 sr)^{-1}}{\sqrt{E_q (s cm^2 sr)^{-1}}} \cdot \qquad (3.5)$$

When ranges are given in meters, detector sizes in micrometers, and fluxes in terms of square centimeters, units must be included in the calculation. A check shows that the seconds cancel nicely, the areas and ranges in meters are fine, but we have both centimeters and micrometers to convert. This problem is resolved if the detector area is converted to centimeters and the radiance is in terms of square meters. Then the units are consistent, and the result can be found in terms of the photon incidance on the detector. The SNR as a function of E_q is illustrated in Fig. 3.5, and shows that the value must be less than 2×10^9 to obtain an SNR of 10. Although an SNR of 10 is low, it might be attainable. Such an SNR will give good values of detection probability and false alarm rates, as described later in Chapter 6.

$\varepsilon := 1 \quad \tau_a := 1 \quad \text{то} := .5 \quad \eta := .8 \quad B := 8 \qquad E_q := 10^9, 1.1 \cdot 10^9 .. 10^{10}$

$c := 2.99793 \cdot 10^{10} \quad h := 6.6256 \cdot 10^{-34} \quad c_2 := 1.4338 \quad \lambda_1 := 0.0003 \quad \lambda_2 := 0.0015 \qquad T := 273$

$R := 10^9 \quad x := 50 \quad D_0 := 10 \quad F := 3 \quad f := F \cdot D_0 \quad \alpha := 10^{-3} \quad a := f \cdot \alpha \quad f = 30$

$A_d := a^2 \qquad A_s := x^2$

$A_0 := \dfrac{\pi \cdot D_0^2}{4} \qquad a = 0.03$

$Lq := \displaystyle\int_{\lambda_1}^{\lambda_2} \dfrac{2 \cdot c}{\lambda^4 \left(e^{\frac{c_2}{\lambda \cdot T}} - 1\right)} d\lambda \qquad SNR(E_q) := \left(\varepsilon \cdot \tau_a \cdot \text{то} \cdot \sqrt{\eta \cdot \dfrac{A_s \cdot A_0}{R^2 \cdot \sqrt{A_d \cdot B}}}\right) \cdot \dfrac{Lq}{\sqrt{E_q}}$

$Lq = 2.685 \cdot 10^{17} \qquad\qquad SNR(10^9) = 8.786$

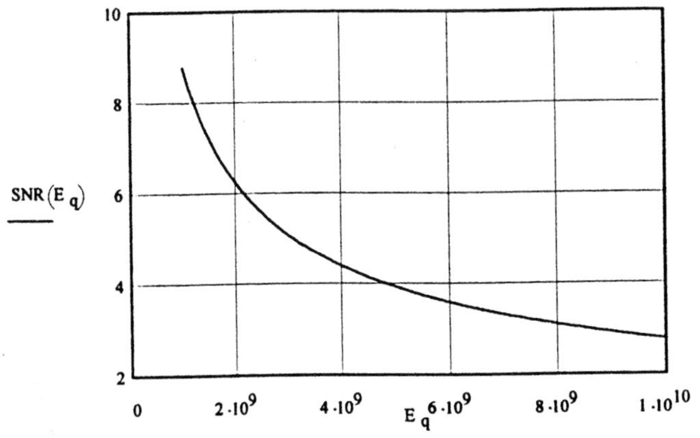

Figure 3.5. SNR as a function of photon incidance for the 3- to 5-μm band.

The essence of the problem now is to evaluate the detector photon incidance and to ensure that it is sufficiently low. If the photon flux density on the detector from all possible sources is less than the value of 2 billion (found above), then the SNR will be more than 10. Can this be attained? We need to examine the sources of such incidance and estimate their values. We also need to determine whether or not a detector with performance as good as this is within reason.

The specific detectivity of a photon detector such as this is given by

$$D^*_{BLIP} = \frac{\lambda}{hc}\sqrt{\frac{\eta}{gE_q}} = \frac{0.015}{6.626x10^{-34}2.9979x10^{10}}\sqrt{\frac{\eta}{4x10^9}} \approx 10^{16} \, . \qquad (3.6)$$

The detectivity calculated in Eq. (3.6) will be cost prohibitive, and may not be available. *The Infrared Handbook* shows a detector that has a detectivity of $4x10^{12}$, a higher incidance of $2.4x10^{13}$, and is still photon limited.[2] *The Infrared and Electro-Optical Systems Handbook*[3] describes a similar detector array with an NEP of <40 aW (attowatt), with an element that can be assessed from the spacing and fill factor to be 68 µm on a side. Plugging in these values results in a specific detectivity of >$1.7x10^{14}$, which is almost good enough and leads us to remain optimistic. We may have to pick up the factor of 100 another way, but we may be able to reach a reasonable solution idealistically.

3.4.2 Sources of detector incidance

Several radiation sources contribute to the total detector incidance. These sources can be classified as in-field, out-of-field, diffuse, and point. The point or diffuse sources can be either in the field or out.

3.4.2.1 The sun

The sun must be classified as an out-of-field source; it is not diffuse--although it is extended. The photon incidance on the aperture can be evaluated from the temperature of the sun and its solid angle,

$$E_{qo} = \int_8^{14} L_q^{BB}(5900,\lambda)d\lambda\Omega = 2.438x10^{16} \, , \qquad (3.7)$$

where L_q is the solar photon radiance and Ω is the solid angle the sun subtends at the earth. The ratio of this rather large value to the required value is the degree of attenuation that must be provided by the baffle tube of the system. Because the sun provides a direct incidance of $2.38x10^{16}$, as determined in Eq. (3.5), and the maximum allowed value is $2x10^9$, the sun cannot shine onto the primary mirror directly, and a baffle must be used that has an attenuation of at least $2.438x10^{16}/2x10^9$ (or about 10^7).

Figure 3.6 is a schematic of the baffle tube. There is usually a specification on how close the sun may come to the field of view of the sensor, because preventing the sun from illuminating any portion (or at least very much) of the primary optic--which in this case is a mirror–is a fundamental concern. The angle θ is equal to the arctangent of *D/L*, the diameter-to-length ratio. Figure 3.7 shows the angle as a function of the *L* over *D* ratio (not *D/L*). The angle α runs from 6 to 60 degrees. The angle Θ is the angle α in radians, and the L/D ratio is given by the reciprocal of the tangent of the angle. If the sun can be no closer than 45 degrees, the tube can have a length equal to the mirror diameter. If, however, the system must view an area closer to the sun, the baffle quickly lengthens. A typical specification is 20 degrees, for which the tube must be 2.7 times the diameter; and for about 10 degrees, the tube is about 5.5 times the diameter. The curve has become almost vertical, and the *L/D* ratio becomes impossible for angles of a few degrees.

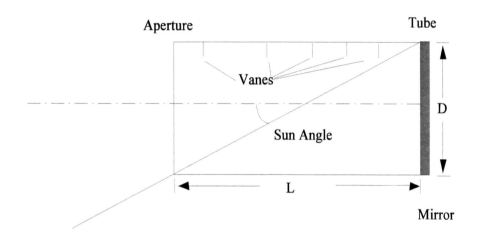

Figure 3-6. Schematic of a baffle tube, showing sun angle, vanes, and mirror.

The overall attenuation is specified in a number of ways. The most useful expression for attenuation is the point source transmittance (PST). It is defined as the ratio of the incidence on the image plane to that on the aperture. This is another area where definitions must be understood carefully. One similar to the above definition of PST is the ratio of *powers* on the two surfaces, and still another is the ratio of either the power or power density *in the input beam* rather than on the aperture. The invidious aspect is that all of these are dimensionless and unitless. The PST is a low number and related inversely to the attenuation. Thus, the requirement becomes a PST of about 10^{-7}.

$$\alpha := 6,7 .. 60$$

$$\Theta(\alpha) := \alpha \cdot \frac{\pi}{180}$$

$$L(\alpha) := \frac{1}{\tan(\Theta(\alpha))}$$

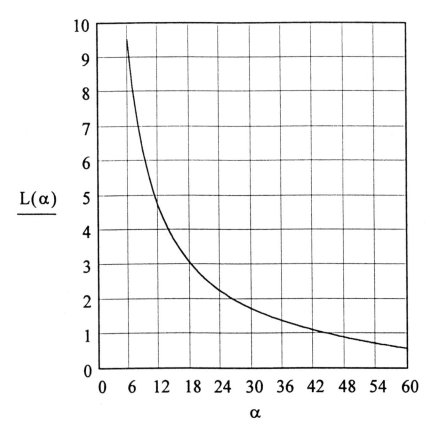

Figure 3.7. The required baffle length L as a function of required off-axis angle (measured in diameters).

To show the nature of the calculation, we can approximate. The procedures for these calculations generally make use of the concept of an equivalent reflectivity for the tube, which factors in the reduced reflectivity that the vanes introduce, and is the reason that they are there. The vane spacing and depth are designed so that there is a reflection from the front of the vane to the tube wall to the back of the preceding vane and then toward the mirror. Also, a direct reflectivity exists from the vane ab to the vane cd. The ratio of the radiance in the output beam to that in the input beam may be considered the effective vane reflectivity. The contribution, per element of circumference, is approximately $(\rho/\pi)^2\ d/s$ and is typically less than 10^{-5}. The contribution from the vane ab to the wall cd and then to the vane cd is about 100 times less. Figure 3.8 shows this geometry.

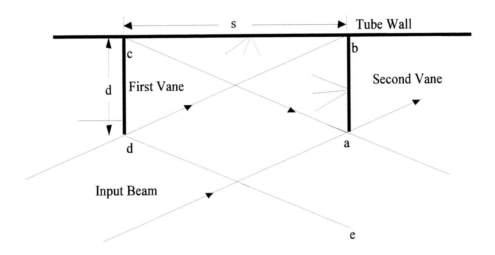

Figure 3.8. Geometry for calculating the effective baffle-wall reflectivity.

The design of the tube can be carried out geometrically. Figure 3.9 shows the tube and the mirror with a line showing the angle at which the sun may sit. The first step is to construct line a-b, which is determined by the allowed solar angle. Next, the depth of the baffle is chosen, the length of a vane. Then construct the line that runs along the vane tips. Then draw line e-f from the bottom vane-tip line to the intersection of c-d and the top vane-tip line. Then draw line c-d, which is parallel to a-b and intersects the vane-tip line and e-f. Draw the first vane from the wall to the intersection point. (That's one!) Now draw line j which is parallel to a-b and meets *e-f* at the baffle wall. Draw *e-g* to the intersection of *j* with the vane-tip line. Repeat the process to get enough vanes. Inspection shows at least two reflections from the vanes and wall before the radiation gets to the mirror. This design allows the use of the effective baffle reflection, described above.

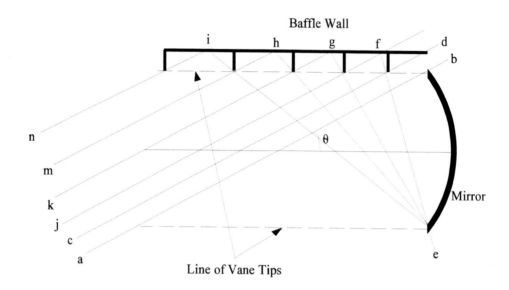

Figure 3.9. Construction of a baffle tube.

Calculation of the PST can be quite difficult and, luckily, several commercial programs are available to do this. The best of these are APART and ASAP. Approximations have been published.[4] Figure 3.10 is a representative plot of a PST.

The technique that the Zeros used in WWII seems to still apply: fly out of the sun.

$$\Theta := 1, 10.01 .. 50 \quad n := 6 \qquad PST(\Theta) := \frac{1}{\Theta^n}$$

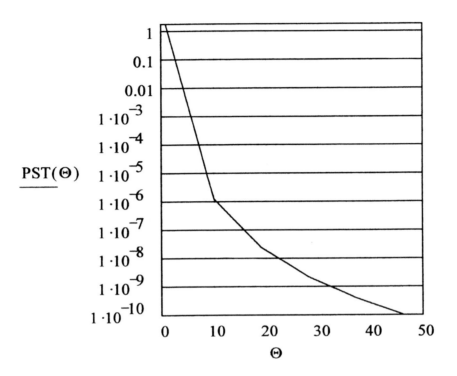

Figure 3.10. A representative PST.

3.4.2.2 The moon

The light from the moon is a reflection of the sun and a contribution from its own emission. The calculation is shown in Fig. 3.11. The first line initializes the speed of light c, the second radiation constant c_2, and the lower and upper wavelengths in centimeters. The second line sets the reflectivity of the moon as lambertian with a hemispherical reflectivity of $0.1/\pi$, the emissivity as the complement of that, and the solid angle of the sun and the moon, Ω_s and Ω_m. The sun subtends 33 arcmin; the moon, 30 arcmin. The sun has a temperature of 5900 K and the sunlit moon 370 K.[5] The photance from the sun is calculated as L_{q1} in the spectral band. That of the sunlit moon, L_{q2}, uses a temperature of 370 K. The dark moon has a photance, L_{q3}, related to a temperature of 120 K. The incidances at the surface of the Earth are indicated by the E_q values. The first, E_{q1}, is the reflected sunlight: the solar photance

$$c := 2.9975 \cdot 10^{10} \qquad c_2 := 1.4388 \qquad \lambda_1 := 0.0003 \qquad \lambda_2 := 0.0015$$

$$\rho := \frac{0.1}{\pi} \qquad \varepsilon := 1 - \rho \qquad \Omega_s := \frac{\pi}{4} \cdot \left(\frac{33}{60} \cdot \frac{\pi}{180}\right)^2 \qquad \Omega_m := \frac{\pi}{4} \cdot \left(\frac{30}{60} \cdot \frac{\pi}{180}\right)^2$$

$$T := 5900$$

$$L_{q1} := \int_{\lambda_1}^{\lambda_2} \frac{2 \cdot c}{\lambda^4 \cdot \left(e^{\frac{c_2}{\lambda \cdot T}} - 1\right)} d\lambda$$

$$E_{q1} := L_{q1} \cdot \rho \cdot \Omega_s \cdot \Omega_m$$
$$L_{q1} = 9.813 \cdot 10^{20}$$

$$T := 370$$

$$L_{q2} := \int_{\lambda_1}^{\lambda_2} \frac{2 \cdot c}{\lambda^4 \cdot \left(e^{\frac{c_2}{\lambda \cdot T}} - 1\right)} d\lambda$$

$$E_{q2} := L_{q2} \cdot \varepsilon \cdot \Omega_m$$

$$T := 120$$

$$L_{q3} := \int_{\lambda_1}^{\lambda_2} \frac{2 \cdot c}{\lambda^4 \cdot \left(e^{\frac{c_2}{\lambda \cdot T}} - 1\right)} d\lambda$$

$$E_{q3} := L_{q3} \cdot \varepsilon \cdot \Omega_m$$

$$E_q := E_{q1} + E_{q2}$$

$$E_{q1} = 1.352 \cdot 10^{11} \qquad E_{q2} = 6.311 \cdot 10^{13} \qquad E_{q3} = 5.57 \cdot 10^{10}$$

$$E_q = 6.324 \cdot 10^{13}$$

Figure 3.11. Radiant photon incidance from the moon.

times the solid angle that the moon subtends at the sun times the reflectivity of the moon times the solid angle that the moon subtends at the Earth. The second, E_{q2}, is the blackbody photance of the sunlit moon times the emissivity, ε, times the moon's solid angle. The third, E_{q3}, is the same for the dark side. The main contribution for this spectral region-- somewhat surprisingly--is that of the sunlit lunar emission. The total incidance from this sunlit moon is 6.324×10^{13}, which requires a baffle attenuation of about 6×10^{4}. The dark moon has a temperature of 120 K and no solar reflection; it provides less incidance, but still a significant value of 5.57×10^{10}.

3.4.2.3 Self radiation

The background is very low, and the incidance on the detector must be very low, so the optical system itself, which is above zero degrees, may contribute too much incidance on the detector. Figure 3.12 shows a typical three-mirror system as a set of three lenses. The calculation can be made as:

$$E_q = \varepsilon_1 L_q^{BB}(T_1)\Omega_1 + \tau_1 \varepsilon_2 L_q^{BB}(T_2)\Omega_2 + \tau_1 \tau_2 \varepsilon_3 L_q^{BB}(T_3)\Omega_3 , \qquad (3.8)$$

where E_q is the detector photon incidance, ε is the emissivity, τ is the transmittance of each element (for mirrors it is the reflectivity), T is the temperature, and Ω is the solid angle that the element subtends at the detector in the optical space of the detector. This means that the image of the second lens by the first must be used, and the image of the third by both the second and the first. The counting goes from right to left, as Fig. 3.12 shows. In many cases, this can be simplified. In particular, if all the elements are at the same temperature, the radiance is the same in each term. If element one is the aperture stop, the solid angles are all alike. Then

$$E_q = (\varepsilon_1 + \tau_1 \varepsilon_2 + \tau_1 \tau_2 \varepsilon_3) L_q^{BB}(T)\Omega . \qquad (3.9)$$

For the three-mirror system,

$$E_q = \left[(1 - \rho_1) + \rho_1(1 - \rho_2) + \rho_1 \rho_2(1 - \rho_3)\right] L_q^{BB}(T)\Omega . \qquad (3.10)$$

If the mirrors all have the same reflectivity, then

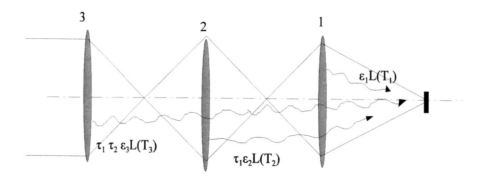

Figure 3.12. Self radiation from a three-element system.

$$E_q = (1-\rho)\left[1+\rho+\rho^2\right]L_q^{BB}(T)\Omega \ . \tag{3.11}$$

This simpler form is often applicable, and it will be used for the rest of our analysis. The more complicated situations are left for the reader (as they usually are for the student).

Figure 3.13 is a plot of the photon incidance as a function of component temperature. An optical system with a speed of $F/3$ was assumed. The result is that the components, for this spectral band, will have to be at about 60 K. The first line sets the values for the speed of light and second radiation constant, c and c_2, the domain for T from 80 down to 50, the optical speed $F=3$, and component reflectivity, ρ at 0.95.

3.4.2.4 Zodiacal Radiance

This source of background is caused by the emission of dust particles distributed throughout the solar system. Data given in *The Infrared Handbook* are summarized in Table 3.1.

$$c_2 := 1.4388 \quad c := 2.997 \cdot 10^{10} \quad T := 80, 79 .. 50 \qquad F := 3 \qquad \rho := .95$$

$$\Omega := \frac{\pi}{4 \cdot F^2}$$

$$L_q(T) := \int_{.0003}^{0.0015} \frac{2 \cdot c}{\lambda^4 \cdot \left(e^{\frac{c_2}{\lambda \cdot T}} - 1\right)} \, d\lambda$$

$$E_q(T) := (1 - \rho) \cdot \left(1 + \rho + \rho^2\right) \cdot L_q(T) \cdot \Omega$$

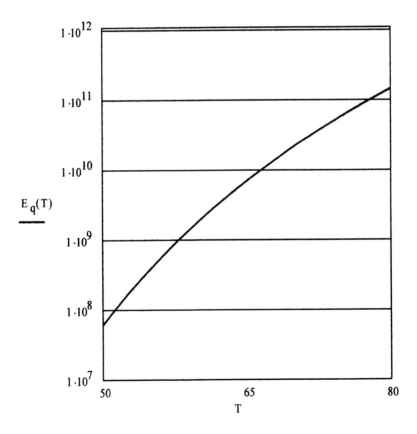

Figure 3.13. System self radiation.

Table 3.1. Zodiacal radiance.		
Spectral Band	Spectral Radiance	Spectral Photance
μm	$Wcm^{-2}sr^{-1} \mu m^{-1}$	$s^{-1}cm^{-2}sr^{-1}\mu m^{-1}$
5-6	$3x10^{-11}$	$8.3x10^{8}$
12-14	$6x10^{-11}$	$16.6x10^{8}$
16-23	$2.5x10^{-11}$	$6.92x10^{8}$

We can take the middle line and calculate, assuming constancy, that the in-band photance would be approximately 10^{10} $s^{-1}cm^{-2}sr^{-1}$. This is the radiance and not the incidance. The incidance can be calculated by multiplying by the optical-speed solid angle or approximately 0.08, which means the incidance is about $8x10^{8}$. These values are for the ecliptic plane, a plane 160 degrees from the sun, and they are upper limits for this plane. The values increase toward the sun and decrease away from it. The Zodiac is acceptable.

3.4.2.5. Diffuse background radiance from extragalactic sources

The Infrared Handbook gives both an equation and a table for a diffuse extragalactic background. Table 3.2 is printed here, but the equation is omitted because this background is not significant.

Table 3.2. Diffuse extragalactic background.							
λ [μm]	3	5	10	15			
L_λ [$wcm^{-2}sr^{-1}\mu m^{-1}$]x$10^{-14}$	11.0	7.2	3.8	2.6	2.0	1.7	1.4

3.4.2.6 Point sources

Point sources are nastier than the diffuse sources discussed above, because they mimic the target. Problems arise when these sources are above the NEP of the system. Charts are presented in both handbook references cited previously for the density of such sources for different spectral regions and angular areas of the sky. The significant parts of these tables are reproduced in Table 3.3, but first we need to know the system NEP, or noise-equivalent-flux density (NEFD). The NEP is the power on the aperture that provides an SNR equal to one. We have, until now, made all our calculations in terms of photons, and this is appropriate for systems that use photon detectors. Unfortunately the tables are given in terms of power. Thus one or the other must be converted.

The NEFD can be approximated by returning to the basic SNR equation, setting the SNR equal to one, and solving for the power:

$$SNR = \frac{D^* P}{\sqrt{A_d B}} \, . \tag{3.12}$$

Then

$$NEFD = \frac{\sqrt{A_d B}}{D^* A_o} \approx \frac{0.001 x 3}{1000 x 10^{16}} = 3 x 10^{-22} \, . \tag{3.13}$$

Table 3.3, excerpted from tables in *The Infrared Handbook*, gives the point sources per square degree above this value of flux density for several spectral bands.

Table 3.3. Point sources per square degree above the NEFD.							
Band [μm]	0	10	20	30	45	60	90
4.2	1,700,000	320	220	150	83	51	35
11	97,000	110	110	4.7	2.8	2.3	2.1
19.8	4.1+3	3.4+1	5.6	2.0	7.9	4.7	3.4

The data vary little with galactic longitude, and they are actually for a flux density of 10^{-20} rather then the required NEFD, but that is as low as they go. Two conclusions can be drawn here. One is to stay away from the plane of the ecliptic at zero latitude. The other is to go to longer wavelengths. Although it may be possible to avoid the ecliptic plane (or not get too close to it), there will still be many point sources detected. In the best case--the lower right of the table--there will be 3.4 detections of natural sources in every square degree. The field of view was determined to be approximately 6x360 degrees or 2160 square degrees--and it is circular. So there will be at least 7344 false alarms in every scan. The two approaches to the solution of this very severe problem are trajectory analysis and spectral discrimination.

It is clear that the trajectory of a celestial body will be quite different from that of the ice cube, and it is fortunate that we used a relatively fast scan so that we could obtain a trajectory.

Spectral discrimination can be accomplished because the celestial bodies, mostly stars, have temperatures in excess of 20,000 K; a simple two-band ratio will perform well as a discrimination technique.

3.5 OPTICS

The system requires a 0.1-rad linear field of view with 0.1-mrad resolution. The usual equations may be applied. Spherical aberration is given by

$$\beta_{SA} = \frac{1}{128F^3} = \frac{1}{128x3^3} = 0.289mr , \qquad (3.14)$$

Although that's a little high, it could be fixed by decreasing the speed with the use of an F/4.3 system. Because these are point sources, the radiometry remains unaffected but the system will be longer. Or we could use an asphere.

The coma is given by

$$\beta_{CA} = \frac{\theta}{16F^2} = 0.\frac{05}{16x3^3} = 0.35[mrad] . \qquad (3.15)$$

This also is a little large. The required speed now is 5.6; it is getting a little long.

The astigmatism is given by

$$\beta_{AA} = \frac{\theta^2}{2F} = \frac{0.05^2}{2x3} = 0.42[mrad] , \qquad (3.16)$$

and the required F/number is 12.5.

This leads me to consider a Schmidt system with no corrector. It has no coma or astigmatism, just spherical aberration, distortion, and curvature of field. It would have to be F/ 4.3. The radius of curvature would be 142 cm and the extreme detectors would be only out of focus if a flat FPA is used. There are pros and cons with the use of a Schmidt. Disadvantages include the curved field and that it must be twice as long as the focal length. Another, for this application, is that it is an obscured system. An obscuration in front of the primary increases the susceptibility to scatter. Advantages are that the primary is a simple sphere, and is the only optical element necessary. That means the system can be warmer than if there are three elements and it is easy to align.

There are other possible choices. Jones[6] reports that several systems can perform adequately for this problem on a flat image plane with a speed of F/1.25: a Schwarzschild, Schmidt, or Maksutov. The latter two require refractive correctors; the Schwarzschild is all-reflective, can be used as an eccentric pupil, and therefore has no obscuration. The conic surfaces take care of the spherical aberration. The Schwarzschild would be my choice.

3.6 THE STEP BACK

The first iteration used a set of assumptions and corrected a few along the way. Now is the time to review and to evaluate. The system can detect the ice cube at a range of 10 Mm, using an array of 1000 very sensitive detector elements in a spectral band from 3 to 15 μm in an eccentric-pupil, all-reflective, unobscured F/1.25, 33-cm-diameter Schwarzschild optical system that rotates a full 360 degrees in 10 seconds. However, at least two bands must be used to discriminate against point targets. The baffle, to prevent the sun from illuminating the primary closer than 20 degrees, must be almost 1 m long and provide an attenuation of 10^7.

Do we really need that range? Can the spectral band be reduced from 15 μm to make both the detector array more practicable and the mirror diameter smaller? Should we move up the short-wave limit to avoid solar flux? Do we need 0.1-mrad position uncertainty? Is the full field the correct size? Can the whole system be made smaller and more economical?

Some of the answers to these questions are in the hands of those who develop the requirements. Some devolve to the infrared system designer. They should talk together so that they can jointly determine the tradeoffs that will provide a superior system.

3.7 TRADEOFFS

If the required range can be reduced, good things happen. An ICBM midcourse trajectory lasts about 20 min. Half of it is from apogee on down, 10 min, 600 s. That gives time for 60 looks, based on the 10-s rotation rate that we have assumed. So detection can be accomplished as late as the second half of the trajectory and, depending on the sensor location, the range could be a little as 1 Mm. This reduces the required sensitivity by a factor of 100, and allows 10,000 times the photon incidance on the detectors--other things being constant. Of course, other things are almost never constant.

Figure 3.14 shows that little target radiance is lost when the spectral band is decreased from 15 μm to 10 μm. Again, the first line sets the speed of light, the second radiation constant, the temperature of the ice cube, and the domain of the longwave limit. From Fig. 3.15, which is exactly the same format, the background flux, characteristic of the sun, can be reduced appreciably by increasing the short-wave limit to about 6 μm. Then the advantage decreases.

$$c := 2.99758 10^{10} \qquad c_2 := 1.4388 \qquad T := 273 \qquad \lambda_2 := .001, .0011 .. .0020$$

$$L_q(\lambda_2) := \int_{.0003}^{\lambda_2} \frac{2 \cdot c}{\lambda^4 \cdot \left(e^{\frac{c_2}{\lambda \cdot T}} - 1\right)} d\lambda$$

$$L_q(.001) = 1.662 \cdot 10^{11} \qquad\qquad L_q(.0015) = 5.154 \cdot 10^{11}$$

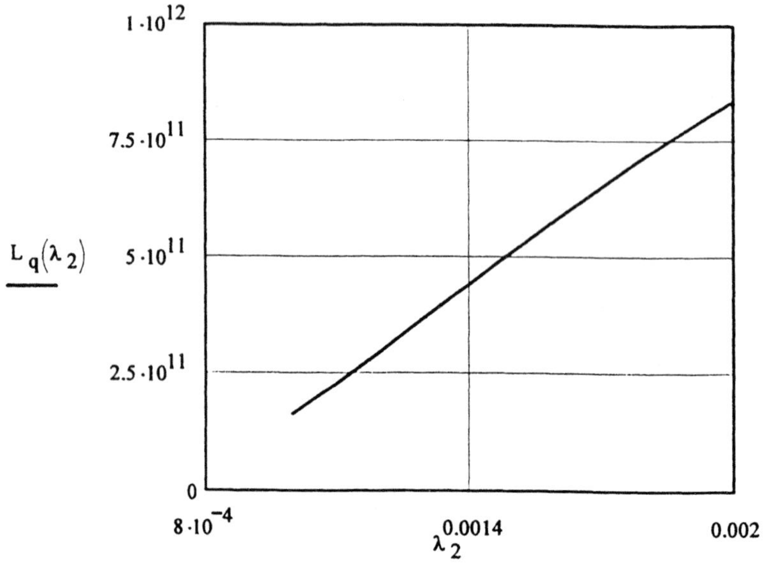

Figure 3.14. Target photance as a function of longwave limit.

$$c := 2.9975810^{10} \qquad c_2 := 1.4388 \qquad T := 5900$$

$$\lambda_1 := 0.0003, 0.00031 .. 0.0008 \qquad\qquad \lambda_2 := 0.001$$

$$L_q(\lambda_1) := \int_{\lambda_1}^{\lambda_2} \frac{2 \cdot c}{\lambda^4 \cdot \left(e^{\frac{c_2}{\lambda \cdot T}} - 1\right)} \, d\lambda$$

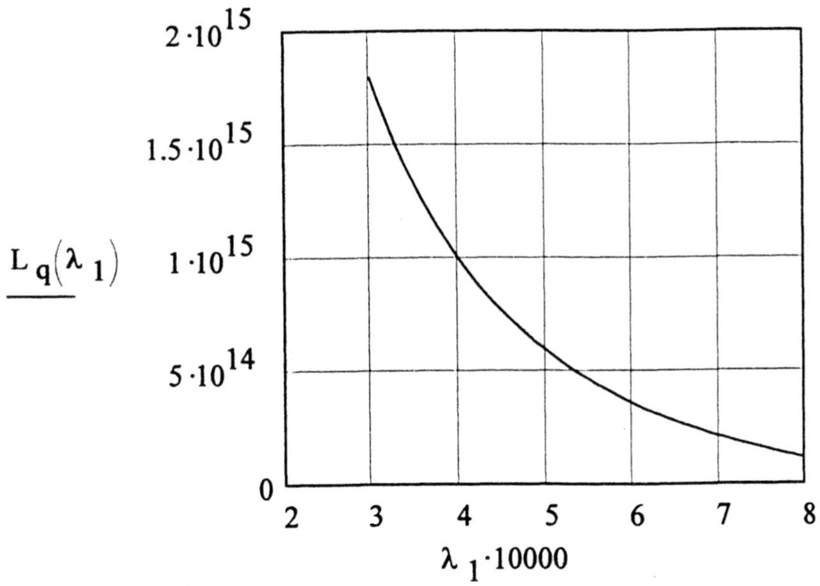

Figure 3.15. Solar background photance as a function of shortwave limit.

Figure 3.16 shows the SNR for the system as a function of the shortwave limit. This rather busy figure might benefit from a line-by-line description of the steps. The first line sets the values for the speed of light, c, the second radiation constant, c_2, the emissivity of the ice cube, ϵ, the atmospheric transmission, τ_a, the optics transmission, τ_o, the detector quantum efficiency, and the equivalent noise bandwidth, B. The second line sets the longwave limit λ_2 at 10 μm and the domain for the shortwave limit from 1 to 10 μm, the range R, the PST, the size of the side

$c := 2.99793 \cdot 10^{10}$ $c_2 := 1.4338$ $\varepsilon := 1$ $\tau_a := 1$ $\tau_0 := .5$ $\eta := .8$ $B := 8$

$\lambda_2 := 0.001$ $\lambda_1 := 0.0001, 0.00011 .. 0.001$ $pst := 10^{-7}$ $R := 10^9$ $x := 50$ $\alpha := 1 \cdot 10^{-3}$

$D_0 := 10$ $F := 3$ $f := F \cdot D_0$ $a := f \cdot \alpha$ $A_s := x^2$ $A_o := \dfrac{\pi \cdot D_0^{\ 2}}{4}$ $A_d := a^2$ $\Omega := \dfrac{\pi}{4} \cdot \left(\dfrac{33}{60} \cdot \dfrac{\pi}{180} \right)^2$

$T := 273$

$$L_{qcube}(\lambda_1) := \int_{\lambda_1}^{\lambda_2} \frac{2 \cdot c}{\lambda^4 \left(e^{\frac{c_2}{\lambda \cdot T}} - 1 \right)} d\lambda$$

$T := 5900$

$$L_{qsun}(\lambda_1) := \int_{\lambda_1}^{\lambda_2} \frac{2 \cdot c}{\lambda^4 \left(e^{\frac{c_2}{\lambda \cdot T}} - 1 \right)} d\lambda$$

$T := 370$

$$L_q(\lambda_1) := \int_{\lambda_1}^{\lambda_2} \frac{2 \cdot c}{\lambda^4 \left(e^{\frac{c_2}{\lambda \cdot T}} - 1 \right)} d\lambda$$

$$E_{qsun}(\lambda_1) := L_{qsun}(\lambda_1) \cdot \Omega$$

$$E_{qmoon}(\lambda_1) := L_q(\lambda_1) \cdot \Omega + 0.9 \cdot E_{qsun}(\lambda_1) \cdot \Omega$$

$$E_q(\lambda_1) := pst \cdot \left(E_{qsun}(\lambda_1) + E_{qmoon}(\lambda_1) \right)$$

$$SNR(\lambda_1) := \varepsilon \cdot \tau_a \cdot \tau_0 \cdot \sqrt{\eta} \cdot \frac{A_s \cdot A_o}{R^2 \cdot \sqrt{A_d \cdot B}} \cdot \frac{L_{qcube}(\lambda_1)}{\sqrt{E_q(\lambda_1)}}$$

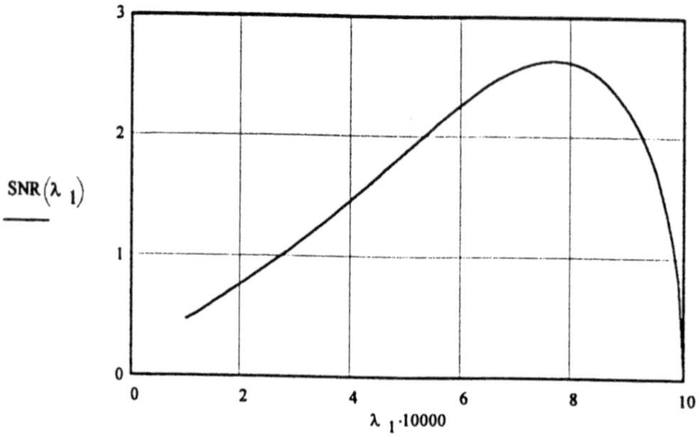

Figure 3.16. SNR as a function of short wavelength limit.

of the cube, x, and the angular resolution of the system, α, at 1 mrad. The third line sets the diameter D_o at 10 cm, the optical speed F at 3, and calculates the optical area A_o, the source area A_s, the focal length f, and the linear dimension of the detector. The fourth line calculates the detector area A_d, the solid angle of the sun, and sets the PST at 10^{-7}. Then, in the fifth line, T is set to 273 K and the photance of the ice cube is calculated by the integral. T is then set at 5900 K, and the integral calculates the photance from the sun. In the next line, T is set to 370 K and the integral calculates the emission from the moon. Then E_{qsun} is calculated as the solar photance times the solar solid angle. The incidence from the moon is calculated as the lunar emission plus reflected sunlight. Finally, the SNR is calculated as a function of the shortwave limit. Figure 3.17 shows the same thing for the longwave limit. In both cases the PST value was assumed to be 10^{-7}. Figure 3.18 shows tradeoffs that result from changing the required range. The left-hand figure shows the required background level on the detector, while the right-hand curve shows the flux density as a function of system temperature. When the range is reduced to 10^8 cm (1 Mm), the flux level requires a system temperature of about 90 K instead of about 45 K. This factor of two can be important in the cryogenic design, but the detection range might not be reduced. Table 3.4 represents my design summation.

Table 3.4. Summary of the campout infrared system		
Property	**Units**	**Value**
Spectrum	μm	6 - 12
Field of Regard	deg	360 x 60
Resolution	mrad	1
Range	km	2000
Optics Type		Eccentric Schwarzschild
Aperture Diameter	cm	2.928
Mirror Diameter	cm	6
Baffle Length	cm	8.193
Overall Length	cm	10
Optics Temperature	K	45-90
Detector Size	μm	87.84
Detector Number	--	1000
Array Size	cm	8.784
Rotation Rate	cps	0.1

$$\varepsilon := 1 \qquad \tau_a := 1 \qquad \tau_0 := .5 \qquad \eta := .8 \qquad B := 8 \qquad \lambda_1 := 0.0007$$

$$c := 2.99793 \cdot 10^{10} \qquad h := 6.6256 \cdot 10^{-34} \qquad c_2 := 1.4338 \qquad \rho := 10^9 \qquad x := 50$$

$$\lambda_2 := 0.001, 0.0011 .. 0.0025 \qquad T := 273 \qquad \qquad \alpha := \frac{x}{\rho} \qquad \alpha = 5 \cdot 10^{-8}$$

$$\alpha := 1 \cdot 10^{-3}$$

$$D_0(\lambda_2) := \frac{2.44\,\lambda_2}{\alpha} \qquad\qquad D_0 := 10 \qquad F := 3 \qquad f := F \cdot D_0 \qquad a := f \cdot \alpha \qquad a = 0.03$$

$$A_0 := \frac{\pi \cdot D_0^{\,2}}{4} \qquad A_s := x^2 \qquad A_d := a^2 \qquad \Omega := \frac{\pi}{4} \cdot \left(\frac{33}{60} \cdot \frac{\pi}{180} \right)^2$$

$$T := 5900$$

$$L_q(\lambda_2) := \int_{\lambda_1}^{\lambda_2} \frac{2 \cdot c}{\lambda^4 \left(e^{\frac{c_2}{\lambda \cdot T}} - 1 \right)} \, d\lambda \qquad\qquad L_q(\lambda_2) := \int_{\lambda_1}^{\lambda_2} \frac{2 \cdot c}{\lambda^4 \left(e^{\frac{c_2}{\lambda \cdot T}} - 1 \right)} \, d\lambda$$

$$T := 370$$

$$L_q(\lambda_2) := \int_{\lambda_1}^{\lambda_2} \frac{2 \cdot c}{\lambda^4 \left(e^{\frac{c_2}{\lambda \cdot T}} - 1 \right)} \, d\lambda \qquad\qquad E_{qsun}(\lambda_2) := L_q(\lambda_2) \cdot \Omega$$

$$E_{qmoon}(\lambda_2) := L_q(\lambda_2) \cdot \Omega + 0.9 \cdot E_{qsun}(\lambda_2) \cdot \Omega$$

$$pst := 10^{-7} \qquad\qquad E_q(\lambda_2) := pst \cdot \left[\left[\left(E_{qsun}(\lambda_2) \right) + E_{qmoon}(\lambda_2) \right] \right]$$

$$SNR(\lambda_2) := \varepsilon \cdot \tau_a \cdot \tau_0 \cdot \sqrt{\eta} \cdot \frac{A_s \cdot A_0}{\rho^2 \cdot \sqrt{A_d \cdot B}} \cdot \frac{L_q(\lambda_2)}{\sqrt{E_q(\lambda_2)}}$$

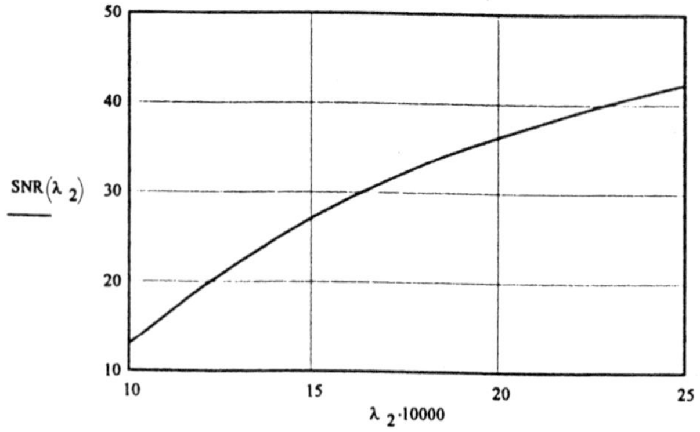

Figure 3.17. SNR as a function of longwave limit.

$\varepsilon := 1 \quad \tau_a := 1 \quad \text{то} := .5 \quad \eta := .8 \quad B := 8 \quad x := 50 \quad \rho := 10^7, 1.1 \cdot 10^7 .. 10^9$

$c := 2.99793 \cdot 10^{10} \quad h := 6.6256 \cdot 10^{-34} \quad c_2 := 1.4338 \quad \lambda_1 := 0.0007 \quad \lambda_2 := 0.0012$

$\alpha := 10^{-3} \quad D_o := \dfrac{2.44 \cdot \lambda_2}{\alpha} \quad F := 3 \quad f := F \cdot D_o \quad a := f \cdot \alpha \quad a = 8.784 \cdot 10^{-3}$

$A_d := a^2 \quad A_s := x^2$

$A_o := \dfrac{\pi \cdot D_o^2}{4} \quad \Omega_{sun} := \dfrac{\pi}{4} \cdot \left(\dfrac{33}{60} \cdot \dfrac{\pi}{180}\right)^2$

$T := 273 \quad L_{qcube} := \displaystyle\int_{\lambda_1}^{\lambda_2} \dfrac{2 \cdot c}{\lambda^4 \left(e^{\frac{c_2}{\lambda \cdot T}} - 1\right)} d\lambda \qquad T := 5900 \quad L_{qsun} := \displaystyle\int_{\lambda_1}^{\lambda_2} \dfrac{2 \cdot c}{\lambda^4 \left(e^{\frac{c_2}{\lambda \cdot T}} - 1\right)} d\lambda$

$r := .9$

$atten := 10^{-7}$

$T := 370 \qquad L_{qmoon} := \displaystyle\int_{\lambda_1}^{\lambda_2} \dfrac{2 \cdot c}{\lambda^4 \left(e^{\frac{c_2}{\lambda \cdot T}} - 1\right)} d\lambda$

$E_{qsun} := L_{qsun} \cdot \Omega_{sun}$

$E_{qmoon} := E_{qsun} + 0.03 \cdot L_{qmoon} \cdot \Omega_{sun}$

$SNR := 10$

$\Omega_{sys} := \dfrac{\pi}{4 \cdot F^2} \qquad E_q := atten \cdot \left(E_{qsun} + E_{qmoon}\right)$

$E_q = 2.082 \cdot 10^9$

$E_q(\rho) := \left[\left(\varepsilon \cdot \tau_a \cdot \text{то} \cdot \sqrt{\eta} \cdot \dfrac{A_s \cdot A_o}{\rho^2 \cdot \sqrt{A_d \cdot B}}\right) \cdot \dfrac{L_{qcube}}{SNR}\right]^2 \qquad T := 50, 51 .. 250$

$L_{qsys}(T) := \displaystyle\int_{\lambda_1}^{\lambda_2} \dfrac{2 \cdot c}{\lambda^4 \left(e^{\frac{c_2}{\lambda \cdot T}} - 1\right)} d\lambda$

$E_{qsys}(T) := (1 - r) \cdot \left(1 + r + r^2\right) \cdot L_{qsys}(T) \cdot \Omega_{sys}$

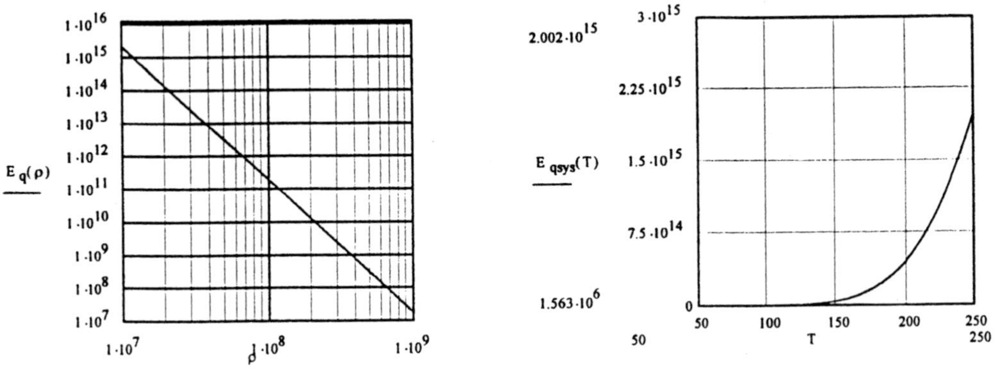

Figure 3.18. Range tradeoffs.

3.8 REFERENCES

[1] W. Wolfe and G. Zissis, *The Infrared Handbook*, The Environmental Research Institute of Michigan (1978); also available from SPIE.

[2] Ibid., pgs 11-93.

[3] *The IR and Electro-Optical Systems Handbook, Vol. III*, p. 269.

[4] A. W. Greynolds, "Formulas for estimating stray-radiation levels in well-baffled optical systems," Proc. SPIE **257**, 39 (1980); J. A. Bamberg, "Stray light analysis with the HP-41C/CV calculator," Proc. SPIE **384**, 29 (1981).

[5] D. R. Lide, ed., *Handbook of Chemistry and Physics*, 73rd edition, CRC Press (1993).

[6] L. Jones, Chapter 18 in *Handbook of Optics, Vol. II*, M. Bass, E. Van Stryland, D. Williams, and W. Wolfe, eds., Mc Graw Hill (1995).

CHAPTER 4
AN INFRARED NIGHT-DRIVING SYSTEM

The military has long investigated the feasibility, utility, and cost of night-driving systems that do not use visible light, or any other illumination. Early systems often used near-infrared headlights with filters that reject the visible. The drivers used Snooperscope viewers. More recently, passive systems that rely on the radiation emitted from the scene have been developed and employed. Of late, the commercial automobile industry has become interested in the prospects[1] of such technology.

Several advantages are inherent in a passive infrared night-driving system. One is that the infrared view is generally better than the view one gets with headlights. Another is that the infrared system is not "blinded" by the headlights of the oncoming car. Still another is that the infrared system does better imaging through light fog and haze. It is my opinion that the infrared system should be included as an aid in driving under adverse conditions--low illumination, haze, and fog.

This chapter delineates the design process for such a system.

4.1 GEOMETRY

The considerations are: range, resolution, and field of view. The assumption is that the imagery will be real time, that is, at normal television rates of 30 frames per second. The range is determined by the stopping distance. Northwestern University Traffic Institute statistics[2] indicate the stopping distances as a function of automobile speed shown in Table 4.1. This range of distances is for different vehicles and speeds and for dry, average pavement. The stopping distance varies with road texture, ambient temperature, tire condition and type, surface wetness, snow or ice, and assorted things on the road like sand, dirt and leaves. For this project we narrow down to the passenger car and light truck. We can assume that the driver slows to 55 mph or less, and that the surface is wet. So, with a certain arbitrariness, that should not be used for the actual design, I choose initially a distance twice the dry road, 55 mph value of 291 ft, which is approximately 582 ft or 180 m.

The resolution is determined by the required number of pixels across the minimum dimension of the chosen object, which in this case is a man (or woman). A typical man is about 2 m high and 1/2 m wide. Thus, the minimum dimension is 0.5 m. The Johnson criterion[3] for recognition of shape requires that there be 8.0 ±

1.6 pixels across the minimum dimension for recognition. This is not recognition of who it is but recognition that it is a who. Because one does not need to know that it is a human, but only that it is a human-sized object, I will reduce this to 5. Thus, the resolution element is 50 cm divided by 5, and therefore 10 cm divided by 180 m, or 0.55 mrad, about 0.5 mrad. The geometry is shown in Fig. 4.1.

Table 4.1. Stopping distances for dry pavement.				
Vehicle	Speed [mph]	Reaction Distance [ft]	Braking Distance [ft]	Total Distance [ft]
Passenger	15	17	14	31
	35	39	78	117
	55	61	230	291
Light Truck	15	17	17	34
	35	39	92	131
	55	61	275	336
Heavy Truck	15	17	22	39
	35	39	125	164
	55	61	310	371
Semi	15	17	29	46
	35	39	160	199
	55	61	390	451

The field of view can be determined in many ways. One is to assume that it should be the field that one sees through a windshield. Somewhat surprisingly, this is about 30 degrees high by 90 degrees wide (about 0.5 rad by 1.5 rad).

Several options are available. What to do? Can it be done with an array? Is a scanner necessary? Should it be a photodetector or a thermal detector? Design is difficult because it is a little like picking the ripest vegetables in the super market when you don't even know which vegetable to pick.

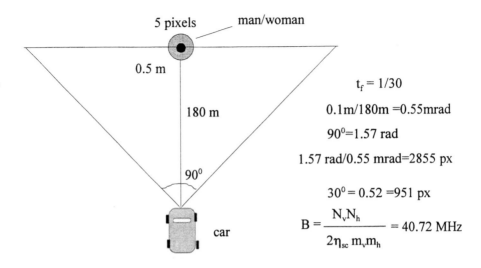

Figure 4.1. Geometry of the night driving system.

4.2 SCANNER DESIGNS

We will take the geometry outlined above and attempt a design solution with a scanner. Many scanners meet our field-of-view and resolution requirements, and potentially have the sensitivity required. They include either the MWIR or LWIR bands, parallel scan, TDI or hybrid scan, PtSi and InSb antimonide detectors for the MWIR, and HgCdTe for the LWIR. We also have the choice of different optical forms, both refractive and reflective, or even catadioptric, keeping in mind that our solution must be affordable. (So, as we reach these forks in the road, we will do as Yogi Berra advised, and take them.)

4.2.1 Geometry

The geometry, as described above, is coverage of a field of 90 degrees wide by 30 degrees high with a resolution of 0.5 mrad, that is (approximately) 1.5 rad by 0.5 rad with 0.5 mrad, 3000 x 1000 pixels. This geometry is quite demanding, and may need to be relaxed by either increasing the resolution angle or decreasing the total field or both.

4.2.2 Dynamics

The field of view must be covered in 1/30 s for real-time imaging. With 100% scan efficiency and a single detector, the time constant would be about 10 ns. That is not possible, and shouldn't be considered for more than the same amount of time. The field could be scanned with a vertical array of 1000 elements, and that would increase the dwell time to 10 μs. This is possible, but presumptuous. The bandwidth would then be 50 kHz. The array could consist of mercury-cadmium- telluride detectors operated at about 70 K to obtain photon-limited detectivity. These numbers are refined in Fig. 4.2.

Figure 4.2 has a somewhat different form than previous Mathcad figure presentations. The first line sets the values for the speed of light, the first radiation constant, and the wavelength limits for the LWIR. It also sets the domain for the temperature from 250 K to 500 K. The next line sets the horizontal and vertical fields of view, Θ_h and Θ_v, at 90 and 30 degrees and converts them to radian measure. It sets the linear pixel size, x, at 0.5 m divided by 7, based on the Johnson criterion. The range, R, is 180 m and the resolution angles, α and β, are assumed equal and calculated as x/R. The next line calculates the number of pixels in each direction, N_h and N_v, sets the frame time at 1/30 s, specifies the number of horizontal and vertical detectors both at 1 and sets the scan efficiency at 1. Then, the next line calculates the dwell time and the bandwidth and displays these values. Because we assume a single detector, the dwell time is found to be in the ns range. Therefore, an array must be used. I chose a linear vertical array that covers the field, so that m_v was set to 1300, as shown in the next line. There are actually 1319 vertical pixels, as shown by $N_v =1.319 \times 10^3$, but the approximation is sufficient. There are $N_h = 3.958 \times 10^3$ horizontal pixels and the scan efficiency is set at $2/\pi$, characteristic of a resonant scanner. Then, as shown in the next line, the dwell time is about 13 μs and the bandwidth is 38.37 kHz. The integrals are used in the NETD calculation, using the idealized equation, and the plot is made of the NETD as a function of temperature--for the domain chosen above.

4.2.3 Sensitivity

The sensitivity can be evaluated on an optimistic basis with the idealistic equation for the LWIR,

$$NETD = \sqrt{B} \left[\frac{L_q}{\lambda \frac{\partial L_q}{\partial T}} \right] = 0.029K \ .$$

(4.1)

$c := 2.9975 \cdot 10^{10}$ $c_2 := 1.4388$ $\lambda_1 := 0.0008$ $\lambda_2 := 0.0012$ $T := 250, 251 .. 500$

$\Theta_h := 90 \cdot \dfrac{\pi}{180}$ $\Theta_v := 30 \cdot \dfrac{\pi}{180}$ $x := \dfrac{.5}{7}$ $R := 180$ $\alpha := \dfrac{x}{R}$ $\beta := \alpha$

$N_h := \dfrac{\Theta_h}{\alpha}$ $N_v := \dfrac{\Theta_v}{\beta}$ $t_f := \dfrac{1}{30}$ $m_v := 1$ $m_h := 1$ $\eta_{sc} := 1$

$t_d := t_f \dfrac{m_v \cdot m_h}{\eta_{sc} \cdot N_v \cdot N_h}$ $B := \dfrac{1}{2 \cdot t_d}$ $B = 7.834 \cdot 10^7$ $t_d = 6.382 \cdot 10^{-9}$

$m_v := 1300$ $N_v = 1.319 \cdot 10^3$ $N_h = 3.958 \cdot 10^3$ $\eta_{sc} := \dfrac{2}{\pi}$

$t_d := t_f \dfrac{m_v \cdot m_h}{\eta_{sc} \cdot N_v \cdot N_h}$ $B := \dfrac{1}{2 \cdot t_d}$ $B = 3.837 \cdot 10^4$ $t_d = 1.303 \cdot 10^{-5}$

$$L_q(T) := \int_{\lambda_1}^{\lambda_2} \frac{2 \cdot c}{\lambda^4 \cdot \left(e^{\frac{c_2}{\lambda \cdot T}} - 1\right)} d\lambda$$

$$dL_q(T) := \int_{\lambda_1}^{\lambda_2} \frac{\frac{c_2}{\lambda \cdot T} \cdot e^{\frac{c_2}{\lambda \cdot T}}}{e^{\frac{c_2}{\lambda \cdot T}} - 1} \cdot \frac{1}{T} \cdot \frac{2 \cdot c}{\lambda^4 \cdot \left(e^{\frac{c_2}{\lambda \cdot T}} - 1\right)} d\lambda$$

$$NETD(T) := \frac{\sqrt{B} \cdot \sqrt{L_q(T)}}{\lambda_2 \cdot dL_q(T)}$$ $NETD(300) = 0.023$

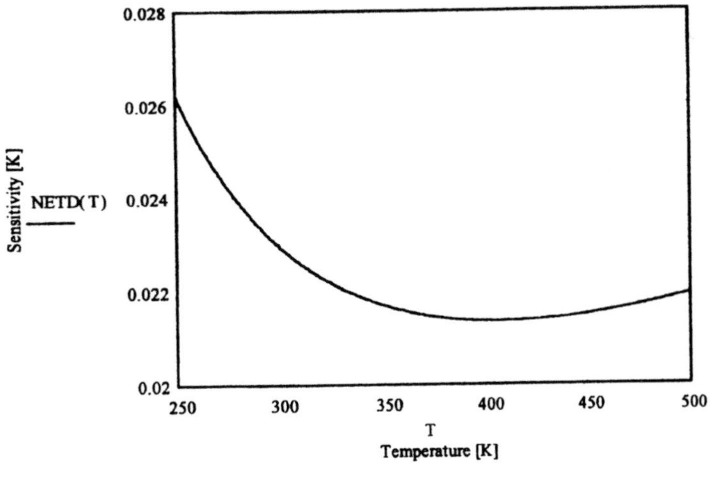

Figure 4.2. Geometry, dynamics, and sensitivity of a night-vision scanner system.

The value of about 0.023 K (at 300 K) is good. Of course, the NETD will be somewhat higher than this when the efficiency factors are included. These include cold-shielding, quantum and scan efficiencies, and atmospheric and optical transmissions. It is clear, however, that even with these included, the NETD will be at least adequate, about twice as high. The dynamics and sensitivity are calculated in Fig. 4.2. As a check, the specific detectivity is calculated for these conditions in Fig. 4.3. The first line is the same as in Fig. 4.2. The second line inserts the value for Planck's constant, sets the domain of the optical speed, F, and calculates the solid angle of the optical system. In the next line, the integral gives the photance the product of the photance and the solid angle gives the incidance, and the final equation is the expression for the photon-limited detectivity D*$_{BLIP}$. The values are in the reasonable range of 10^{11}. For the 3- to 6-μm range, the NETD at 300 K is 0.087 K (obtained by temporary substitution in the Mathcad formulation).

4.3 OPTICAL DESIGN

The optical system must cover 0.5 rad (30 degrees) by 0.5 mrad, because it is a scanner. We will start with the canonical F/3 system for the simple optical resolution evaluation. We recall that the field angle should be half of the total field of the linear array, which is about 0.25 rad (actually 0.262 rad).

Spherical aberration is given by

$$\beta_{SA} = \frac{1}{128F^3} = 0.289 \; [mrad] \; . \tag{4.2}$$

That is OK. The coma is

$$\beta_{CA} = \frac{\theta}{16F^2} = \frac{0.262}{144} = 1.8 \; [mrad] \; . \tag{4.3}$$

The astigmatism is

$$\beta_{AA} = \frac{\theta^2}{2F} = \frac{0.262^2}{6} = 11 \; [mrad] \; . \tag{4.4}$$

$c := 2.9975 \cdot 10^{10}$ $c_2 := 1.4388$ $\lambda_1 := 0.0008$ $\lambda_2 := 0.0014$ $T := 300$ $\eta := 0.8$

$h := 6.626 \cdot 10^{-34}$ $F := 1, 2 .. 6$ $\Omega(F) := \dfrac{\pi}{4 \cdot F^2}$

$$L_q := \int_{\lambda_1}^{\lambda_2} \dfrac{2 \cdot c}{\lambda^4 \cdot \left(e^{\frac{c_2}{\lambda \cdot T}} - 1\right)} d\lambda$$ $E_q(F) := L_q \cdot \Omega(F)$ $D_{BLIP}(F) := \dfrac{\lambda_2}{h \cdot c} \cdot \sqrt{\dfrac{\eta}{g \cdot E_q(F)}}$

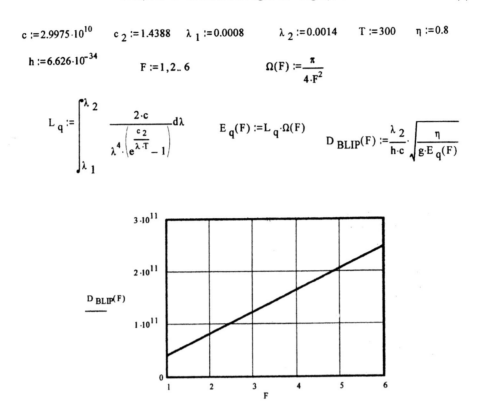

Figure 4.3. Theoretical D* for the night-vision photon detector.

The aberrations are shown in Fig. 4.4 as a function of the field of view and for an optical speed of 3. The first line sets this speed as F=3 and calculates the half-field angle in radians, the angular resolution as five pixels across a man of 0.5 m width at a distance of 180 m, and sets the domain of the half-field angle from 0 to Θ_m. Because this is a scanner, the off-axis angle is half of the vertical field that is scanned by the linear array. Both astigmatism and coma are too large at the edge of the field. Two options are immediately apparent. One is to use half as many detectors in the linear array, cutting the field in half (or in fourth or sixth). This would bring the coma close and the astigmatism only about twice what it ought to be. From the Mathcad presentation, a four-pass scan will allow the aberrations to be small enough. In such a case, the dwell time will be reduced by a factor of four and the NETD increased by a factor of two. That will do it. The other is to use a correctorless Schmidt, which eliminates the off-axis aberrations, and the spherical aberration is already satisfactory. That will make the system somewhat longer. We could try a multi-element lens, but we must keep the costs low.

$$F := 3 \qquad \Theta_m := 15 \cdot \frac{\pi}{180} \qquad \alpha := \frac{.5}{5} \cdot \frac{1}{180} \qquad \Theta := 0, .01 .. \Theta_m$$

$$\beta_{SA}(\Theta) := \frac{1}{128 \cdot F^3} \qquad \beta_{CA}(\Theta) := \frac{\Theta}{16 \cdot F^2} \qquad \beta_{AA}(\Theta) := \frac{\Theta^2}{2 \cdot F}$$

$$\beta_{SA}(.262) = 2.894 \cdot 10^{-4} \qquad \Theta_m = 0.262 \qquad \alpha = 5.556 \cdot 10^{-4}$$

$$\beta_{CA}(\Theta_m) = 1.818 \cdot 10^{-3} \qquad \qquad \beta_{AA}(\Theta_m) = 0.011$$

$$\beta_{CA}\left(\frac{\Theta_m}{2}\right) = 9.09 \cdot 10^{-4} \qquad \qquad \beta_{AA}\left(\frac{\Theta_m}{2}\right) = 2.856 \cdot 10^{-3}$$

$$\beta_{CA}\left(\frac{\Theta_m}{4}\right) = 4.545 \cdot 10^{-4} \qquad \qquad \beta_{AA}\left(\frac{\Theta_m}{4}\right) = 7.139 \cdot 10^{-4}$$

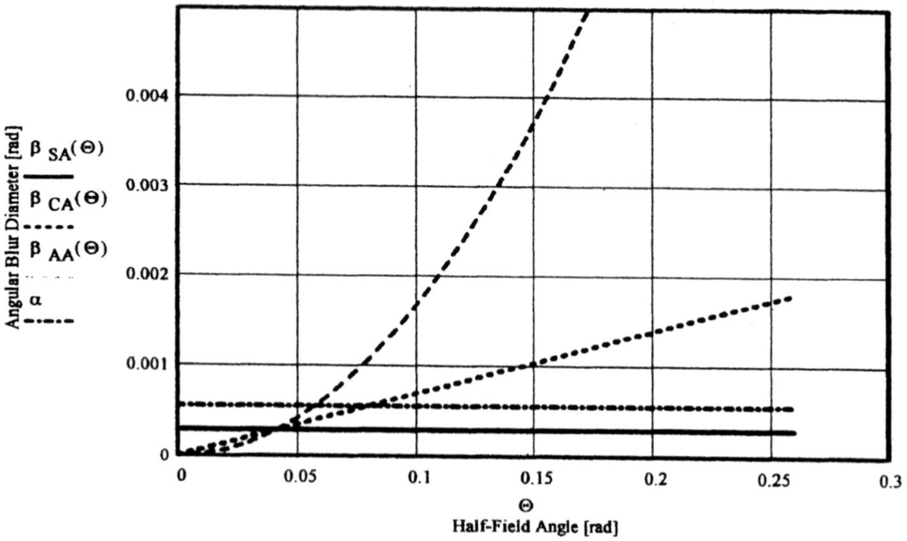

Figure 4.4. Required resolution and third-order aberrations.

The scanner can be an oscillating mirror on a galvanometer. We can assume a reimaging lens and therefore a cold-shielding efficiency of 100%. For mercury-cadmium-telluride detectors the quantum efficiency is about 0.8. The scan must be made in 1/30 s. The scan efficiency for a resonant scanner is $2/\pi$. Figure 4.5 shows the data for such a system. Atmospheric transmission must be calculated for the rather short path, but one that might have light fog, mist and other absorbants. The night-vision device must work on cold clear nights, but also under foggy and misty conditions. The atmospheric transmission under these rather nasty conditions, as calculated by PCModWin (the drizzle model), is 0.88. The NETD is still a magnificent 2 mK!

Table 4.2 is a summary of four designs, not all of which have been explicitly calculated here.

Table 4-2. Four night-vision scanners.					
Design	Units	One	Two	Three	Four
Spectrum	μm	3-6	3-6	8-12	8-12
Instant Field	radxmrad	0.066x5.6	0.066x5.6	0.066x5.6	0.066x5.6
Optical Form	--	Schmidt	Lens	Schmidt	Lens
Optics Diameter	cm	3	3	6	
Optics Speed	--	3	3	3	3
Focal Length	cm	9	9	9	9
Overall Length	cm	18	9	18	9
Pixel Size	cm	9	9	9	9
Pixel Subtense	mrad	0.56	0.56	0.56	0.56
Detector	--	PtSi	InSb	HgCdTe	HgCdTe
Array	--	325x1	325	325	325
Element Size	μm	45	45	45	45
Array Size	cmxμm	1.46 x 45	1.46 x 45	1.46 x 45	1.46 x 45
NETD @ 300 K	K	~1.0	0.174	0.051	0.051

$c := 2.9975 \cdot 10^{10} \quad c_2 := 1.4388 \quad \lambda_1 := 0.0008 \quad \lambda_2 := 0.0012 \quad T := 300 \quad h := 6.626 \cdot 10^{-34}$

$\Theta_m := \dfrac{15}{180} \cdot \pi \quad \Theta := 0, .01 .. \Theta_m \quad m_h := 1024 \quad m_v := m_h \quad \alpha := \dfrac{\Theta_m}{m_h} \quad \beta := \alpha$

$t_i := \dfrac{1}{30} \quad\quad B := \dfrac{4}{2 \cdot t_i} \quad\quad r := 180 \quad\quad a := r \cdot \alpha \quad\quad F := 3 \quad\quad D_o := 6 \quad\quad D_{star} := 10^{11}$

$\varepsilon := 1 \quad\quad \eta_{sc} := 1 \quad\quad \eta := .8 \quad\quad \eta_{cs} := 1 \quad\quad \tau_o := 0.9 \quad\quad \tau_a := 0.88$

$$\beta_D(\Theta) := \frac{2.44 \cdot \lambda_2 \cdot 0.0001}{D_o} \qquad \beta_{SA}(\Theta) := \frac{1}{128 \cdot F^3} \quad \beta_{CA}(\Theta_h) := \frac{\Theta_h}{16 \cdot F^2} \qquad \beta_{AA}(\Theta_h) := \frac{\Theta_h{}^2}{2 \cdot F}$$

$$dL := \int_{\lambda_1}^{\lambda_2} \frac{\dfrac{c_2}{\lambda \cdot T} \cdot e^{\frac{c_2}{\lambda \cdot T}}}{e^{\frac{c_2}{\lambda \cdot T}} - 1} \cdot \frac{1}{T} \cdot \frac{2 \cdot c^2 \cdot h}{\lambda^5 \cdot \left(e^{\frac{c_2}{\lambda \cdot T}} - 1 \right)} \, d\lambda$$

$$NETD := \frac{2}{\pi} \cdot \frac{F}{D_o} \cdot \frac{\sqrt{B}}{\tau_a \cdot \tau_o \cdot \varepsilon} \cdot \frac{1}{\alpha} \cdot \frac{1}{D_{star} \cdot dL}$$

$$NETD = 1.934 \cdot 10^{-3}$$

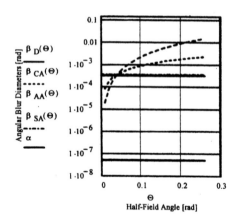

Figure 4.5. Geometry, dynamics, and sensitivity of an HgCdTe night-vision scanning system.

4.4 STARER DESIGN

The starer may not cover the full 90 degrees of horizontal field. About the best we can do is 30 degrees--a circular field of about 30 degrees. We can use a 1024-by-1024 array of platinum-silicide detectors that operates in the 3- to 6-μm region or mercury-cadmium-telluride array that operates in the 8- to 12-μm region with about 512-by-512 detectors. Here we are definitely limited by the technology of the available detector arrays.

4.4.1 Geometry

The geometry to be considered is a field of about 30 degrees (0.5 rad) divided into either 1024 or 512 pixels, yielding a resolution of 0.5 mrad or 1 mrad.

4.4.2 Dynamics

The dynamics calculation is even easier. A starer has a dwell time that is equal to the frame time, that is, 1/30 s.

4.4.3 Sensitivity

The sensitivity equations have all been given in terms of the specific detectivity, and that is all right. However, it makes somewhat more sense to calculate the electrons on the detector from the source and the electronic noise in terms of electrons for a platinum silicide detector system. The NETD can again be found by taking the temperature derivative of the SNR.

The operative equation is

$$NETD = \frac{\sqrt{N_s + N_i}}{\dfrac{\partial N_s}{\partial T}}, \tag{4.5}$$

where the electrons generated in an integration time t_i from the signal, N_s is given by

$$N_s = \epsilon L_q^{BB} \eta \eta_{ff} \tau_o \tau_a \alpha^2 A_o t_i, \tag{4.6}$$

where ϵ is the emissivity of the source, L_q^{BB} is the photonic radiance, η is the detector quantum efficiency, η_{ff} is the fill factor of the array, τ_o and τ_a are the atmospheric and optical transmittances, α is the pixel subtense, A is the optics area, and N_i is the mean square internal electronic noise (in electrons). These are all plugged into the Mathcad expressions in Fig. 4.6. The fact that the NETD hardly

$c := 2.9975 \cdot 10^{10}$ $c_2 := 1.4388$ $\lambda_1 := 0.0003$ $\lambda_2 := 0.0006$ $T := 300$ $h := 6.626 \cdot 10^{-34}$

$\Theta_m := \dfrac{15}{180} \cdot \pi$ $\Theta := 0, .01 .. \Theta_m$ $m_h := 1024$ $m_v := m_h$ $\alpha := \dfrac{\Theta_m}{m_h}$ $\beta := \alpha$

$t_i := \dfrac{1}{30}$ $r := 180$ $a := r \cdot \alpha$ $F := 3$ $D_o := 6$ $A_o := \dfrac{\pi}{4} \cdot D_o^2$

$\varepsilon := 1$ $\eta_{sc} := 1$ $\eta := 0.01$ $\eta_{cs} := 1$ $\eta_{ff} := 0.5$ $\tau_o := 0.9$ $\tau_a := 0.88$

$\beta_D(\Theta) := \dfrac{2.44 \cdot \lambda_2 \cdot 0.0001}{D_o}$ $\beta_{SA}(\Theta) := \dfrac{1}{128 \cdot F^3}$ $\beta_{CA}(\Theta_h) := \dfrac{\Theta_h}{16 \cdot F^2}$ $\beta_{AA}(\Theta_h) := \dfrac{\Theta_h^2}{2 \cdot F}$

$$L_q := \int_{\lambda_1}^{\lambda_2} \dfrac{2 \cdot c}{\lambda^4 \cdot \left(e^{\frac{c_2}{\lambda \cdot T}} - 1\right)} \, d\lambda$$

$$dL_q := \int_{\lambda_1}^{\lambda_2} \dfrac{\frac{c_2}{\lambda \cdot T} \cdot e^{\frac{c_2}{\lambda \cdot T}}}{e^{\frac{c_2}{\lambda \cdot T}} - 1} \cdot \dfrac{1}{T} \cdot \dfrac{2 \cdot c}{\lambda^4 \cdot \left(e^{\frac{c_2}{\lambda \cdot T}} - 1\right)} \, d\lambda$$

$N_s := \eta \cdot \eta_{ff} \cdot \tau_a \cdot \tau_o \cdot \alpha^2 \cdot t_i \cdot A_o \cdot L_q$

$dN_s := \eta \cdot \eta_{ff} \cdot \tau_a \cdot \tau_o \cdot \alpha^2 \cdot t_i \cdot A_o \cdot dL_q$

$NETD(N_i) := \dfrac{\sqrt{N_s + N_i}}{dN_s}$ $N_s = 3\,662 \cdot 10^6$

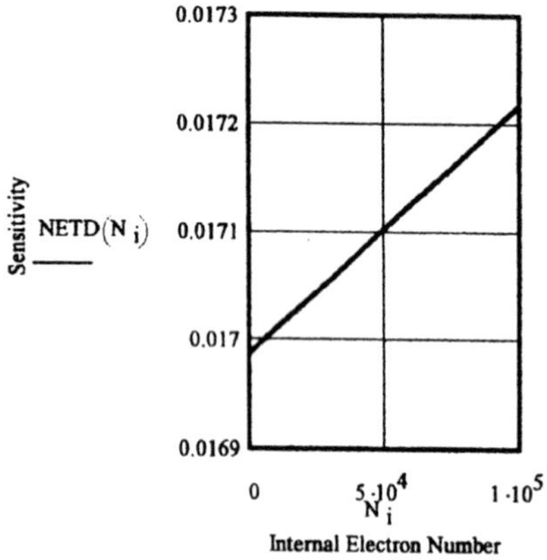

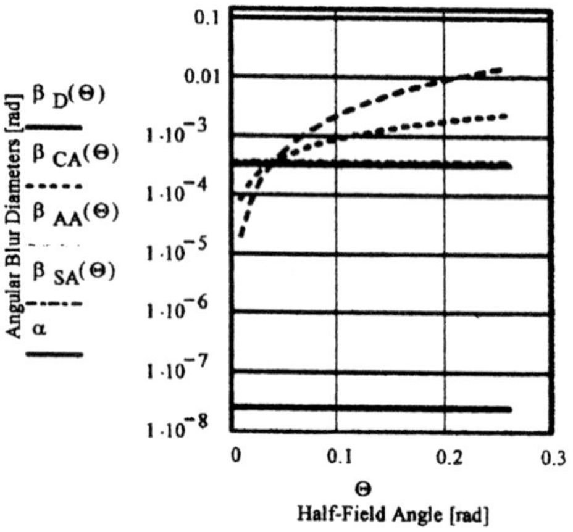

Figure 4.6. Geometry, dynamics, and sensitivity of a PtSi night-vision scanning system.

changes with increasing internal noise shows that the system is essentially background limited, and the value is great. However (and the major reason for a caveat), there are too many electrons. The value is several million, and the well size is about ten times less than that. This can be solved with a switching process that reduces the integration time by a factor of about 10. This will also increase the NETD by about that amount, to 0.17 K. The use of a fill-and-spill technique, that is, dump all the electrons part-way through the integration time and then integrate some more, is a viable system also.

The HgCdTe system clearly meets the requirements, as can be inferred easily by reference to the scanning system that uses a 512 x 4 linear array. The bandwidth will be smaller, and therefore the NETD will also be smaller than that for the scanner. The comparison will devolve to array uniformity and techniques for compensating a lack thereof.

4.4.4 Optics

The problem is the design of a low-cost optical system that will cover a 30-degree field of view with resolution of either 0.5 or 1 mrad. The curves show that the aberrations which violate the requirement are the field aberrations of coma and astigmatism. I am led to a Schmidt system, again.

The LWIR version of this has only 512 detectors on a side, making the optical design easier but with poorer resolution. The sensitivity should be better, and the diffraction-limited diameter should be larger. These results are shown in Fig. 4.7.

4.4.5 Step back

We have found several systems that work: a scanner that covers 90 degrees and two starers, one in the MWIR and the other in the LWIR. The latter showed a sensitivity of about 5 mK at 300K. This leads to the consideration of an array with less sensitivity, notably an uncooled array.

4.4.6 An uncooled system

Sensitivity analysis of a system using an uncooled array must be based on the available properties of that array. These include specific detectivities of between 10^8 and 10^9 and a format of 250 by 350 elements. This arrangement is analyzed in Fig. 4.8. The NETD is less than 100 mK for most reasonable values of detector sensitivity, and at 100 m, the linear resolution is 15 cm, giving 3 to 4 pixels across the waist of an average person, perhaps clothed. The angular resolution is just under 1.5 mrad, which should allow a reasonable optical design (Fig. 4.9). The third-order

$$c := 2.99 \cdot 10^{10} \quad h := 6.626 \cdot 10^{-34} \quad c_1 := 2 \cdot \pi \cdot c^2 \cdot h \cdot 10^{16} \quad c_2 := 14388 \quad T := 300 \quad \lambda_1 := 8 \quad \lambda_2 := 14$$

$$t_f := \frac{1}{30} \quad t_d := t_f \quad B := \frac{1}{2 \cdot t_d} \quad D_o := 6 \quad F := 3 \qquad D_{star} := 2 \cdot 10^8, 2.1 \cdot 10^8 .. 10 \cdot 10^8$$

$$\tau_a := 0.8 \quad \tau_o := .5 \quad \varepsilon := 1 \qquad \Theta := \frac{30}{180} \cdot \pi \quad N_h := 350 \quad \alpha := \frac{\Theta}{N_h}$$

$$dL := c_1 \cdot \int_{\lambda_1}^{\lambda_2} \frac{\frac{c_2}{\lambda \cdot T} \cdot e^{\frac{c_2}{\lambda \cdot T}}}{\lambda^5 \cdot \left(e^{\frac{c_2}{\lambda \cdot T}} - 1\right)^2} \cdot \frac{1}{T} \, d\lambda$$

$$\alpha = 1.496 \cdot 10^{-3}$$

$$R := 0, 1 .. 200$$

$$\Theta = 0.524$$

$$x(R) := \alpha \cdot R$$

$$NETD\left(D_{star}\right) := \frac{4}{\pi} \cdot \frac{F}{D_o} \cdot \frac{\sqrt{B}}{\tau_a \cdot \tau_o \cdot \varepsilon} \cdot \frac{1}{\alpha} \cdot \frac{1}{D_{star} \cdot dL}$$

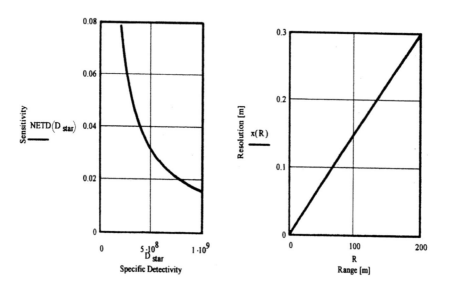

Figure 4.8. Characteristics of the uncooled LWIR system.

aberration approximations show that only astigmatism is a problem, and that it becomes a problem only above 0.1 rad, about 10 degrees. Astigmatism is inversely proportional to the F/number. Thus, to get three times less aberration, the F/number would have to be 9--a little too much. A Schmidt system, or even a pseudo Schmidt, is a reasonable option. Off-axis aberrations disappear when the stop is at the center of curvature, but they are reduced as the stop is moved away from the mirror. We probably still want a single mirror for production purposes. Another consideration, that might be combined with the stop position, is that a certain amount of aberration at the edges of the field may be tolerable.

4.4.7 System choice

The choices are a scanner system that covers 30 degrees by 90 degrees, an MWIR system with 1024 detectors on a side and NETD of about 0.15 at 300 K, an LWIR

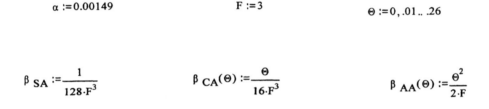

$$\alpha := 0.00149 \qquad\qquad F := 3 \qquad\qquad \Theta := 0, .01 .. .26$$

$$\beta_{SA} := \frac{1}{128 \cdot F^3} \qquad\qquad \beta_{CA}(\Theta) := \frac{\Theta}{16 \cdot F^3} \qquad\qquad \beta_{AA}(\Theta) := \frac{\Theta^2}{2 \cdot F}$$

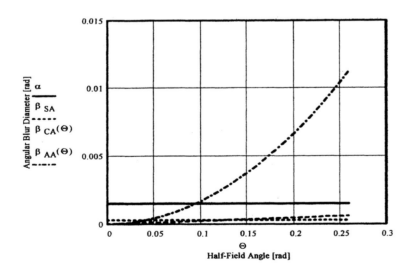

Figure 4.9. The aberrations of the uncooled system.

system with an NETD of 5 mK and 512 x 512 detectors (each with a speed of about F/3 and diameter of 6 cm, but with different resolutions) and an uncooled system with 250-by-350 elements and an NETD of about 30 mK. The choice is based on elimination of cooling, the cost, and the reliability. The latter design bears additional scrutiny. This is a real-time imaging system. It benefits from the integration of frames and the adaptability of the human eye to act as a matched filter. Such systems have come to be described by their MRTD curves, and we will now make that calculation.

4.5 MRTDs

The minimum-resolvable temperature difference (MRTD), is the NETD as a function of spatial frequency, whereas the NETD itself is only for zero spatial frequency, and including the temporal and spatial integration effects of the eye when the system views a four-bar resolution chart with a bar aspect ratio of seven.[4]

We will calculate the spatial frequency responses of the various components of the system as far as we can. Then we will incorporate the integration improvement factors.

4.5.1 MTF

Modulation transfer functions (MTFs), are defined as the modulus of the complex frequency response function of the system components in the spatial-frequency domain. Thecomponent MTFs are multiplied to obtain the system MTF. They consist of the MTF of the optics, the spatial and temporal MTF of the detector, the MTF of the electronics and of the display. Each can be a highly detailed calculation. We will start with the optics.

A diffraction-limited optical system has a response given by[5]

$$MTF(\xi) = \frac{2}{\pi}\left[\cos^{-1}\left(\frac{\lambda\xi}{2NA}\right) - \frac{\lambda\xi}{2NA}\sin\left(\cos^{-1}\left(\frac{\lambda\xi}{2NA}\right)\right)\right]. \qquad (4.7)$$

This can be rewritten in terms of the cutoff frequency, at which point the MTF goes to zero. The cutoff frequency is given by

$$\xi_o = \frac{2NA}{\lambda} = \frac{1}{\lambda F} \, , \tag{4.8}$$

so that

$$MTF(\chi) = \frac{2}{\pi}\left[\cos^{-1}\chi - \chi\sin(\cos^{-1}\chi)\right], \tag{4.9}$$

where χ is the spatial frequency normalized to the cutoff frequency. This is the autoconvolution of the aperture function of the system[6] and is plotted in Fig. 4.10.

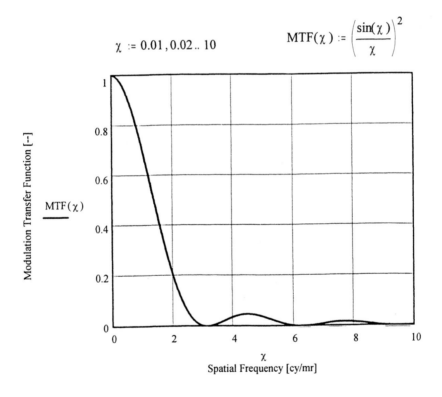

$$\chi := 0.01, 0.02 .. 10 \qquad MTF(\chi) := \left(\frac{\sin(\chi)}{\chi}\right)^2$$

Figure 4.10. The one-dimensional spatial MTF of a square detector.

The spatial MTF of the detector is the Fourier transform of the detector itself. Assuming that the detector has a uniform areal sensitivity and is square, the MTF is a sinc function, as shown in Fig. 4.11. The temporal MTF is the spatially converted temporal response of the detector. Assuming that the detector and its electronics can be characterized by a single time constant, the response is a typical lorentzian, as shown in Fig. 4.12. Other filters can be used, such as Butterworths, that roll off more steeply. A second-order Butterworths is also shown in Fig. 4.12.

$$\chi := 0, .01 .. 1$$

$$MTF(\chi) := \frac{2}{\pi} \cdot (acos(\chi) - \chi \cdot sin(acos(\chi)))$$

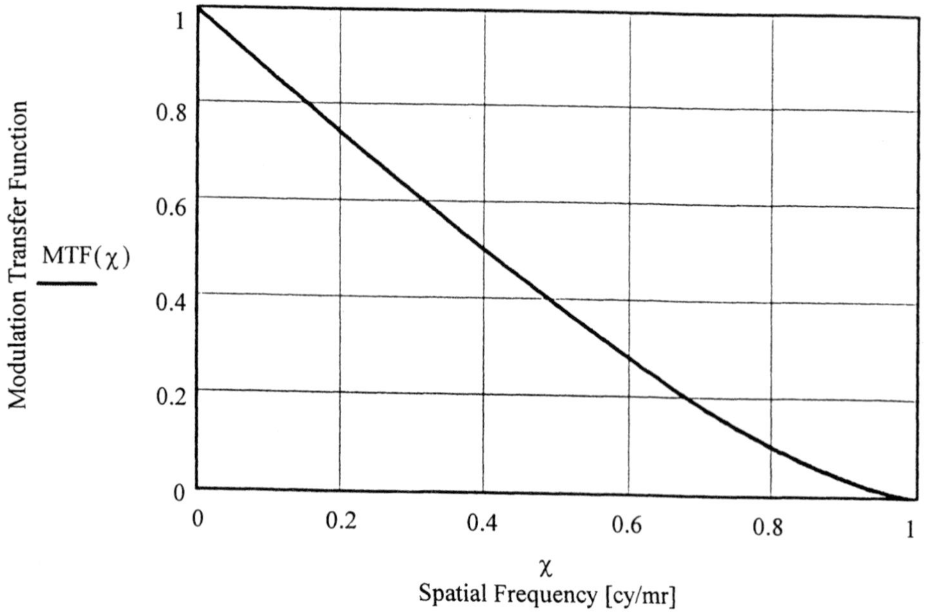

Figure 4.11. The MTF of a diffraction-limited system.

$$\chi := 0.01, 0.02 .. 1$$

$$\text{MTF}_t(\chi) := \frac{1}{1 + (2 \cdot \pi \cdot \chi)^2}$$

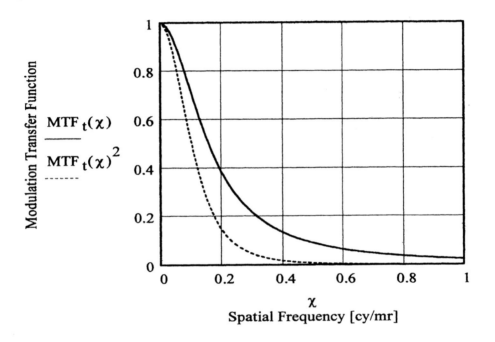

Figure 4.12. The temporal MTF of detectors with different rolloffs.

The electronics can usually be designed so that they do not affect the overall MTF.

The display for this kind of system will almost certainly be an array of LEDs, and therefore have an MTF similar to that of the detector array. A CRT would be characterized by a gaussian.

The MTF of the system is the product of these non-normalized responses, as shown in Fig. 4.13. The non-normalized responses are obtained by considering the real cutoff frequencies. For a diffraction-limited optical system, that is $\xi = 1/\lambda F$. In this case, it is $\xi = D/\lambda = 1/0.014$ mm x 3 = 23.8 cy/mm or, by multiplying by the focal length, 4.28 cy/mrad. The spatial cutoff frequency of the detector is given by the reciprocal of the length of its side. Actually, the length of a side is half a cycle,

$$\chi := 0.01, 0.02.. \; 1$$

$$\text{MTF}_t(\chi) := \frac{1}{1 + (2 \cdot \pi \cdot \chi)^2}$$

$$\text{MTF}_o(\chi) := \frac{2}{\pi} \cdot (\text{acos}(\chi) - \chi \cdot \sin(\text{acos}(\chi)))$$

$$\text{MTF}_d(\chi) := \frac{\sin(\chi)}{\chi}$$

$$\text{MTF}(\chi) := \text{MTF}_o(\chi) \cdot \text{MTF}_d(\chi) \cdot \text{MTF}_t(\chi)$$

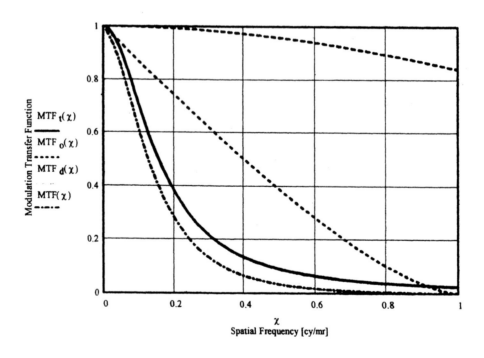

Figure 4.13. The MTFs.

but the fill factor of the array is 50%, so the spatial frequency cutoff is 20 cy/mm or 3.6 cy/mrad. The spatial frequency response of the temporal response of a detector in a scanner is found by the size of the detector and the speed of scan. In the case of a starer, the frame time is important, and the ratio of the frame time to the detector response time is the critical dimension. These values are $1/30 \text{ s} = 0.033$ and 0.001 s or a factor of 33. With so many time constants in the integration time, the MTF is flat. The MRTD is the NETD divided by this frequency response and multiplied by the eye factors, treated below.

4.5.2 Vision factors

Some of these real-time viewing factors actually improve the operation of the system; others take into account the geometry of the standard test target, a four-bar target with a 7:1 aspect ratio.

The first (real) improvement factor arises from the fact that the frame time is 1/30 (0.033) of a second whereas the eye has a time constant of just about 1/5 (0.2) of a second. Thus, six frames are integrated, and the MRTD is lower than the NETD by $\sqrt{6}$ or 2.5.

The second (real) improvement factor is that the eye mysteriously works as a matched filter. This also provides an improvement in the noise of the system.

The third factor arises because the bar chart has a 7:1 aspect ratio. The spatial integration factor perpendicular to the observation direction (in the direction of the bars) provides a gain of $\sqrt{7} = 2.65$. This is only an improvement for observing bar charts.

For a bar target such as this the eye apparently extracts the mean signal level, which is $2/\pi$, the average value of the first harmonic of the square wave. This is not an improvement (more a detriment), but it must be included.

Lloyd[4] combines these factors and some perception information for 90% probability of bar detection to get an equation that he says in his experience, "...adequately predicts the measured MRTD of a detector-noise-limited, artifact-free system..." It is

$$MRTD = \frac{3NETDk\sqrt{\alpha/\alpha_T}}{MTF\sqrt{t_e/t_f}}, \qquad (4.10)$$

where k is the square root of the ratio of the matched-filter bandwidth to the system bandwidth, α is the (across-bar) angular subtense of a pixel, α_T is the lateral subtense of the bar, t_e is the eye integration time, and t_f is the frame time. (I have made some minor substitutions in Eq. (5.53) of Lloyd's[4] previous work). We can assume that the system noise is flat in the frequency band, and obtain that k is just the square root of f_T/B (by direct integration of the sinc function). For a staring system we must take this fraction to be one. When the system views a set of four bars set at the limiting geometric resolution $\sqrt{(\alpha/\alpha_T)}$ is also one. Then the MRTD is equal to the NETD times $3/\sqrt{6} = 1.22$, but divided by the MTF. For our purposes, after all of this, we can assume that, at zero spatial frequency, when the MTF is one, the MRTD equals the NETD. Figure 4.14 shows the MRTD of the uncooled system.

This figure indicates that the NETD is approximately 0.2 K, so that temperature difference of 0.2 K can be sensed with an SNR of 1. Similarly, at the specified spatial frequency of 1 cycle per mrad (reciprocal of two pixels), about 0.3 K can be detected with an SNR of 1. At the higher resolution for recognition, 1.6 cycles per milliradian, the value is 0.4 K, and for an SNR of 5 it is 2 K--a very reasonable temperature difference.

4.6 REMARKS

Because this system is to be used on a car, some installation and presentation issues must be addressed. Where does the sensor head go? It is probably about the size of a portable bar-code scanner, 8 cm in diameter by 10 cm long. It could go high on the windshield, but it cannot go behind the glass window. The windshield could be re-formed to accommodate the sensor head, but that would require that production procedures be totally overhauled. It could be installed near the headlights, but it would get splashed. It could go on the hood near the windshield, but the hood gets hot and would both warm the unit and provide unwanted atmospherics.

The display could be heads up; and that may come. The initial design was a dashboard display via a CRT or LED arrangement. It could be combined with a visual display that can represent red and green stoplights. The human engineering of the display needs considerable investigation, but not here.

4.7 HELICOPTER NIGHT VISION

Advanced versions of helicopters need better night vision. I participated in the design of such devices with McDonnell Douglas and Honeywell. It was done to a large extent the way real-time imagers should be designed--from the inside out. This is an approach my friend Luc Biberman has long espoused: study the requirements of the user in terms of the display luminance, pixel size, number, background, and other psychovisual aspects of the problem; start on the inside and let those decisions drive the sensor.

$c := 2.99 \cdot 10^{10}$ $h := 6.626 \cdot 10^{-34}$ $c_2 := 1.4388$ $\lambda_1 := 0.0008$ $\lambda_2 := 0.0012$ $D_0 := 6$ $F := 3$

$\tau_a := .88$ $\tau_0 := .5$ $\varepsilon := 1$ $T := 300$ $\Theta := \dfrac{30}{180} \cdot \pi$ $\alpha := \dfrac{\Theta}{N_h}$ $D_{star} := 5 \cdot 10^8$ $N_h := 350$

$$dL := 2 \cdot c^2 \cdot h \cdot \int_{\lambda_1}^{\lambda_2} \dfrac{\dfrac{c_2}{\lambda \cdot T} \cdot e^{\frac{c_2}{\lambda \cdot T}}}{\lambda^5 \cdot \left(e^{\frac{c_2}{\lambda \cdot T}} - 1\right)^2} \cdot \dfrac{1}{T} \, d\lambda$$

$t_f := \dfrac{1}{30}$ $t_d := t_f$ $B := \dfrac{1}{2 \cdot t_d}$ $t_e := 0.1$

$$NETD := \dfrac{4}{\pi} \cdot \dfrac{F}{D_0} \cdot \dfrac{\sqrt{B}}{\tau_a \cdot \tau_0 \cdot \varepsilon} \cdot \dfrac{1}{\alpha} \cdot \dfrac{1}{D_{star} \cdot dL}$$

$k := 1$

$\xi := 1, 2 .. 5000$

$\xi_d := \dfrac{D_0}{1.22 \cdot \lambda_2}$ $\chi(\xi) := \dfrac{\xi}{\xi_d}$ $\xi_d = 4.098 \cdot 10^3$ $MTF_s(\xi) := \dfrac{\sin(\chi(\xi))}{\chi(\xi)}$

$$MTF_0(\xi) := \dfrac{2}{\pi} \left(acos(\chi(\xi)) - \chi(\xi) \cdot \sin(acos(\chi(\xi))) \right)$$

$$MTF(\xi) := MTF_0(\xi) \cdot MTF_s(\xi)$$

$$MRTD(\xi) := \dfrac{3 \cdot NETD \cdot k}{MTF(\xi) \cdot \sqrt{\dfrac{t_e}{t_f}}}$$

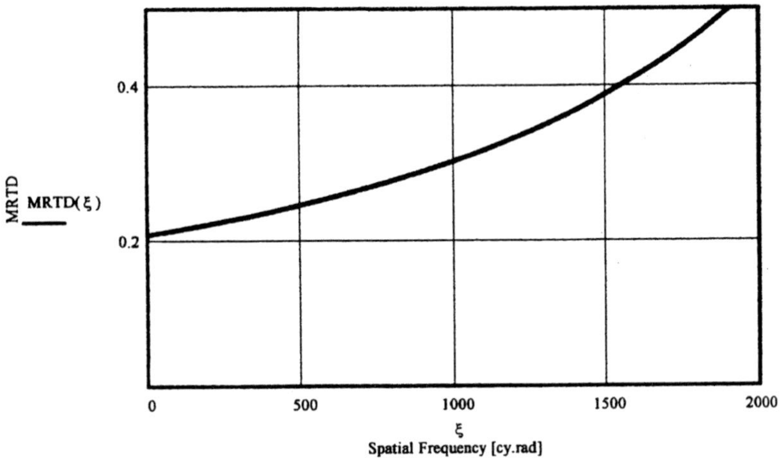

Figure 4.14. MRTD of the uncooled LWIR system.

4.7.1 Psychovisual inputs

McDonnell Douglas provided the psychovisual inputs. They included the full field of regard that was required by the pilot. There were several, dictated by different functions.

> A horizontal field of view of 220 degrees was driven by hover and slow sideways flight maneuvers.
>
> An upward vertical field of 60 degrees is needed for coordinated turns.
>
> A downward vertical field of 45 degrees is required for descent and landing.

The required resolution of the display was dictated by acuity characteristics of the eye. The highest acuity of the eye is in the 0.5-degree central region of the fovea, where it is about 1 arcmin (0.1 arcmin vernier) or 0.3 mrad. The eye has reasonable acuity as much as 10 degrees from the center, where is is about 20/100. Thus, a reasonable full field is about 20 degrees or 0.35 rad. Thus, there would be about 1000 pixels across the field.

The eye has several different movements. The saccadic motions are small, but very necessary for operation. The eye can make larger motions, without head movement, up to about 25 degrees from center. Information at larger angles is considered peripheral, and peripheral cues can help, particularly with sensors and displays that can slew.

Visual acuity is a function of display luminance, although that will not affect the sensor design.

The rate of scan--that is, the frame rate--is a function of the display luminance, or the operator will be bothered by flicker. The combination of these dictates a frame rate of 30 Hz or more and a luminance of 10 foot lamberts.

There are additional and more subtle aspects of the display characteristics, but these are enough to show how the characteristics of the eye and the display should drive the design of the infrared sensor.

4.7.2 Display results

Different methods display this information, but we are interested in how the display characteristics drive the sensor design. From the above, the field should have high resolution over ±10 degrees with lower resolution over a larger field to accept peripheral cues and the frame rate must be at least 30 Hz.

This has been a very brief summary of the study on both psychovisual factors and the displays. The report by Justice et al.[7] should be consulted for detailed information.

4.7.3 Sensor derivatives

So, the sensor should have 1000 pixels across the full field of view of 20 degrees with a resolution of 0.3 mrad. It should also have a poorer resolution over a larger field of view.

My solution for these requirements is a combination of a scanning sensor with a double array of HgCdTe detectors, 512x4, plus two outrigger sensors that consist of an array of 250x350 microbolometers--both of which cover 20 degrees.

4.8 REFERENCES

[1] J. Eaton, "Cadillac will offer option to help with night vision," *The Denver Post*, August 23, 1998.

[2] G. N. Drake, *Survival Behind the Wheel*, Phoenix Books/Publishers (1987).

[3] J. Johnson, "Analysis of image forming systems," Proceedings of Image Intensifier Symposium, US Army Engineering and Development Lab, Ft. Belvoir, VA (1958), available in R. B. Johnson and W. L. Wolfe, *Infrared System Design*, Vol. 513 SPIE Milestone Series (1985); L. M. Biberman, ed., *Perception of Displayed Information*, Plenum (1973); R. Lombardo, Jr., "Target acquisition," Photonics Spectra, p. 123 (July 1998).

[4] J. M. Lloyd, *Thermal Imaging Systems*, Plenum Press (1975).

[5] W. Smith, *Modern Optical Engineering*, McGraw Hill (1990).

[6] J. Goodman, *Introduction to Fourier Optics*, McGraw Hill (1968).

[7] Justice, S., J. Seeman and J. Wasson, Advanced Helicopter Pilotage Final Report, Volume II, McDonnell Douglas Helicopter Company, 5000 McDowell Road, Mesa, AZ 85205-9797, 1992.

CHAPTER 5
THE BROAD OCEAN SURVEILLANCE SYSTEM (BOSS)

The Earth's surface has more water than land, and the monitoring of naval ships and armadas is an important function in global warfare. Ocean surveillance systems have commercial applications as well, and both will be discussed in this chapter. We begin by examining military applications.

5.1 REQUIREMENTS

This hypothetical problem, inspired in part by Tom Clancy's *Red Storm Rising*,[1] is the constant observation of the North Atlantic Ocean, although another area could equally well be considered. The general requirements are that vessels as small as destroyer escorts would be detectable, and the vessels should be observed in a manner that is timely with respect to their speed and maneuverability. Only surface vessels will be considered; submarines are a very different matter. Continuity, detectability, and coverage are the issues.

Several possible approaches to this include a geosynchronous satellite that can sit in position and monitor the entire area in a single frame, the same satellite that covers the area in several frames, and a lower satellite that passes over the area periodically. The airplane is not an option (as it was briefly in the MX problem).

5.2 GEOMETRY

Let us postulate a geosynchronous satellite with a mirror that has a 1-m diameter. Then at 6 μm, the resolution spot is

$$\beta = \frac{2.44\lambda}{D} = \frac{2.44 x 6 x 10^{-6}}{1} \approx 15 \mu rad \tag{5.1}$$

A synchronous satellite orbits at an altitude of 33 Mm, so the spot at nadir for this spectral band is 483 m. For the LWIR the spot will be approximately 3 times this or 1127 m. These are both bigger than a ship, but that will not stop us. These will be the pixel sizes for the two spectral bands in the first design iteration. These are illustrated in Fig. 5.1.

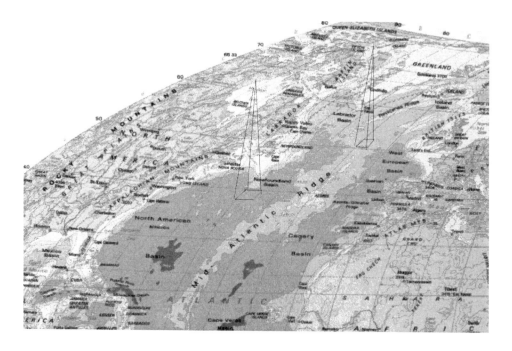

Figure 5.1. The broad-ocean area.

The North Atlantic Ocean is approximately six time zones wide (east-west) and is 2/3 as high (north-south), 4000 by 6000 miles, or 6.4 x 9.6 Mm. This is equivalent to 5678x8518 pixels for the LWIR and 19,870x13,250 for the MWIR. No existing array can do either. So we can look forward to a step-stare or a scanning mode. The step starers must make 185 and 250 subfields, not counting scan efficiencies to cover the full field. It is reasonable to require that the ships move no more than one-tenth of a pixel from frame to frame. This would be 48.3 and 112.7 km. The velocity is about 30 knots, or 48 km/s. This requires 1.1 and 2.6 s respectively for the entire field, or bandwidths of about 100 Hz and 10 Hz. These calculations are shown in Fig. 5.2. The first line of Fig. 5.2 sets the values for X and Y, the linear field sizes of 6.4 and 9.6 Mm, the range R from the satellite to these fields of 33 Mm, the aperture diameter D of 100 cm, and the area of the ship A_{ship} as 300x30 m^2. The next line sets the wavelength for the MWIR approach. The line after has the calculation of α, the diffraction-limited angular subtense, the spot size, x, on the ground, the number N_h and N_v of pixels in the horizontal and vertical directions. Then, in the next line are calculated the number of subfields in each direction and the total number of subfields, N_{fh}, N_{fv} and N_f. The next line shows the velocity, v, converted from knots to meters per second (in two steps) and the time for a field, t_f, the time it takes to move one-tenth of a pixel, and for a dwell, the time

$X := 6.4 \cdot 10^6$ $Y := 9.6 \cdot 10^6$ $R := 33 \cdot 10^6$ $D := 100$ $A_{ship} := 300 \cdot 30$

$$\lambda := 0.0006$$

$\alpha := \dfrac{2.44 \cdot \lambda}{D}$ $x := R \cdot \alpha$ $N_h := \dfrac{X}{x}$ $N_v := \dfrac{Y}{x}$ $x = 483.12$

$N_{fh} := \dfrac{N_h}{1024}$ $N_{fv} := \dfrac{N_v}{1024}$ $N_f := N_{fv} \cdot N_{fh}$ $N_v = 1.987 \cdot 10^4$

$N_f = 251.039$

$v := 30 \dfrac{55}{30}$ $v := v \cdot \dfrac{1000}{3600}$ $t_f := \dfrac{0.1 \cdot x}{v}$ $t_d := \dfrac{t_f}{N_f}$ $B := \dfrac{1}{2 \cdot t_d}$ $N_h = 1.325 \cdot 10^4$

$B = 39.693$

$Ratio := \dfrac{A_{ship}}{x^2}$ $Ratio = 0.039$

$$\lambda := 0.0014$$

$\alpha := \dfrac{2.44 \cdot \lambda}{D}$ $x := R \cdot \alpha$ $N_h := \dfrac{X}{x}$ $N_v := \dfrac{Y}{x}$ $x = 1.127 \cdot 10^3$

$N_h = 5.677 \cdot 10^3$

$N_{fh} := \dfrac{N_h}{1024}$ $N_{fv} := \dfrac{N_v}{1024}$ $N_f := N_{fv} \cdot N_{fh}$ $N_v = 8.516 \cdot 10^3$

$t_f := \dfrac{0.1 \cdot x}{v}$ $t_d := \dfrac{t_f}{N_f}$ $B := \dfrac{1}{2 \cdot t_d}$ $t_f = 3.162$

$N_f = 46.109$

$Ratio := \dfrac{A_{ship}}{x^2}$ $B = 3.125$

$Ratio = 7.082 \cdot 10^{-3}$

Figure 5.2. Geometry and dynamics of the BOSS.

for a subfield, and the bandwidth, one over twice the dwell time. Finally the ratio of a ship area to a pixel area is calculated. On the right are displayed the values of these calculations. The wavelength is set to 14 µm, 0.0015 cm, and the calculations are repeated.

5.3 SENSITIVITY

We again use the idealized equation for both spectral regions.

$$NETD^{-1} = \frac{\lambda_m}{B^{1/2}} \frac{\frac{dL_q}{dT}}{\sqrt{\int Lq}} \, , \tag{5.2}$$

where the photances are in-band values and all efficiencies are one, except for the detector quantum efficiency. The values for the two spectral bands (LWIR and MWIR) are calculated in Fig. 5.3. The first line sets the constants, the wavelengths, the bandwidth, and the temperature domain. The integrals get the photance and photance difference. The next line sets the quantum efficiency for PtSi and calculates the NETD; then it does it for InSb. The process is repeated for the LWIR. Because the ships do not fill the pixels, we must consider the area ratios to obtain equivalent NETD values. Table 5.1 indicates the dimensions and areas of some typical World War II ships.[2]

Today's ships are somewhat larger, but we may take a typical ship as having an area of approximately 9000 sq m, compared to the pixels of 233,289 and 1,270,129 sq m. If, by this assumption, we have forced the enemy to eliminate heavy cruisers, battleships, and aircraft carriers from his armada, then the ratios are 0.007 and 0.039 for the LWIR and MWIR. These are the equivalent NETDs calculated in the figure. Although the LWIR has higher intrinsic sensitivity (lower NETD), it has a larger pixel area, and there is little to choose between the InSb and LWIR systems. They each have an equivalent NETD of just about 0.15 K. Detection of the ships should be easy.

Data indicate that the average ship temperature is at least 10 K above the ocean temperature.[3] The data are not given explicitly but must be inferred from the thermograph images. The ships go to saturation on the image, indicating at least 10 K. We can take this as a reasonable value, but should make some measurements before we do a serious calculation. The sensitivity calculations were made for an ocean that ranges from 250 K to 350 K, and should probably be refined to colder temperatures for the northern Atlantic. If the temperature of the ocean is assumed to be colder, the NETD value rises, but the temperature difference between the ship and the ocean rises too. There is ample trade space and many design opportunities.

$c := 2.9975 \cdot 10^{10}$ $c_2 := 1.4388$ $\lambda_1 := 0.0003$ $\lambda_2 := 0.0006$ $B := 40$ $T := 250, 251 .. 350$

$$L_q(T) := \int_{\lambda_1}^{\lambda_2} \frac{2 \cdot c}{\lambda^4 \cdot \left(e^{\frac{c_2}{\lambda \cdot T}} - 1\right)} d\lambda$$

$$dL_q(T) := \int_{\lambda_1}^{\lambda_2} \frac{\frac{c_2}{\lambda \cdot T} \cdot e^{\frac{c_2}{\lambda \cdot T}}}{e^{\frac{c_2}{\lambda \cdot T}} - 1} \cdot \frac{1}{T} \cdot \frac{2 \cdot c}{\lambda^4 \cdot \left(e^{\frac{c_2}{\lambda \cdot T}} - 1\right)} d\lambda$$

$\eta := 0.01$ $NETD_{PtSi}(T) := \dfrac{\sqrt{B} \cdot \sqrt{L_q(T)}}{0.0039 \cdot \lambda_2 \cdot dL_q(T)} \cdot \dfrac{1}{\sqrt{\eta}}$ $\eta := 0.80$ $NETD_{InSb}(T) := \dfrac{\sqrt{B} \cdot \sqrt{L_q(T)}}{0.039 \cdot \lambda_2 \cdot dL_q(T)} \cdot \dfrac{1}{\sqrt{\eta}}$

$\lambda_1 := 0.0008$ $\lambda_2 := 0.0014$ $B := 3.125$

$$L_q(T) := \int_{\lambda_1}^{\lambda_2} \frac{2 \cdot c}{\lambda^4 \cdot \left(e^{\frac{c_2}{\lambda \cdot T}} - 1\right)} d\lambda$$

$$dL_q(T) := \int_{\lambda_1}^{\lambda_2} \frac{\frac{c_2}{\lambda \cdot T} \cdot e^{\frac{c_2}{\lambda \cdot T}}}{e^{\frac{c_2}{\lambda \cdot T}} - 1} \cdot \frac{1}{T} \cdot \frac{2 \cdot c}{\lambda^4 \cdot \left(e^{\frac{c_2}{\lambda \cdot T}} - 1\right)} d\lambda$$

$$NETD_{LWIR}(T) := \frac{\sqrt{B} \cdot \sqrt{L_q(T)}}{0.007 \cdot \lambda_2 \cdot dL_q(T)} \cdot \frac{1}{\sqrt{\eta}}$$

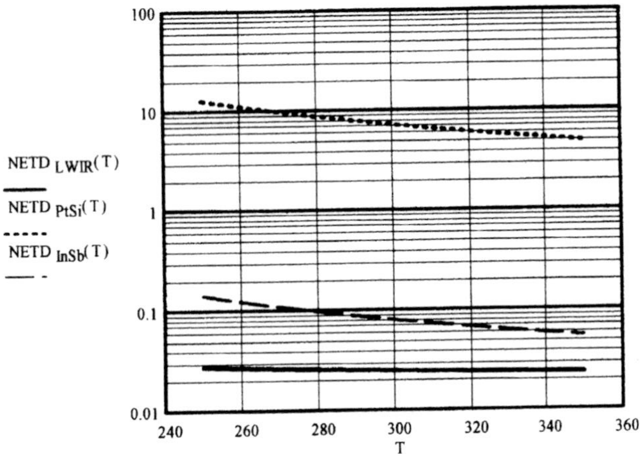

$NETD_{LWIR}(T)$
———
$NETD_{PtSi}(T)$
- - - - -
$NETD_{InSb}(T)$
— ·

Figure 5.3. Idealized NETDs of the BOSS.

Table 5.1. Selected ship dimensions.				
Ship	Units	Length	Width	Area
Carrier	ft (m)	950 (290)	200 (61)	(17690)
Battleship	ft (m)	600 (183)	150 (46)	(8418)
Heavy Cruiser	ft (m)	650 (198)	150 (46)	(9108)
Light Cruiser	ft (m)	600 (183)	100 (30)	(5490)
Destroyer	ft (m)	350 (107)	50 (15)	(1605)

5.4 OPTICS

The resolution has been driven by the available diffraction limits, with an assumed aperture diameter of one meter. Arrays of 512x512 are available for the LWIR and in InSb or 1024x1024 in PtSi for the MWIR. These calculations are illustrated in Fig. 5.4. We have NETD values of about 0.15 K and temperature differences of about 10 K. That gives some margin that may be taken by efficiencies.

I have used trial and error with the Mathcad after my initial guesses. For the MWIR, using PtSi, there are 1024 diffraction-limited elements. One can use an F/5.5 optical system with a 50-cm diameter and the optics turn out to be just a simple parabola. The array is about 8.5 cm square, providing little obscuration (and none if it is an eccentric pupil system). I changed the optical speed and the diameter until I got a system I liked. Note that this changes the area ratio so that the NETD must be recalculated.

The LWIR system is shown in the lower part of Fig. 5.4. Again, I adjusted the optical speed until I got to a system I liked. This is the single parabola of 100 centimeter diameter, retaining the area ratio calculated above.

5.5 STEP BACK

The sensitivities are adequate for the idealized photon-limited photon detectors. All efficiency factors must then be included. The scan efficiency of a step-staring system is probably about 90 %. Cold shielding will be about 50%, and fill factor is also about 50 %. This is considered in Fig. 5.5. The only disadvantage is that cooled detectors are needed, which will shorten the satellite lifetime. Some cooling methods involve closed-cycle refrigerators and even solid cryogens,[4] both of which have a limited lifetime. The weather satellite uses a radiative cooler, which might be a reasonable option. The final option is the solid-state Peltier cooler with no moving parts.

$F := 5.5$ $\lambda := 0.0006$ $D := 50$ $\beta_d := \dfrac{2.44 \cdot \lambda}{D}$ $\Theta_m := 512 \cdot \beta_d$ $\Theta := 0, .001 .. \Theta_m$

$\beta_{SA} := \dfrac{1}{128 \cdot F^3}$ $\beta_{CA}(\Theta) := \dfrac{\Theta}{16 \cdot F^2}$ $\beta_{AA}(\Theta) := \dfrac{\Theta^2}{2 \cdot F}$ $a := F \cdot D \cdot \beta_d$ $1024 \cdot a = 8.245$

$\Theta_m = 0.015$ $a = 8.052 \cdot 10^{-3}$

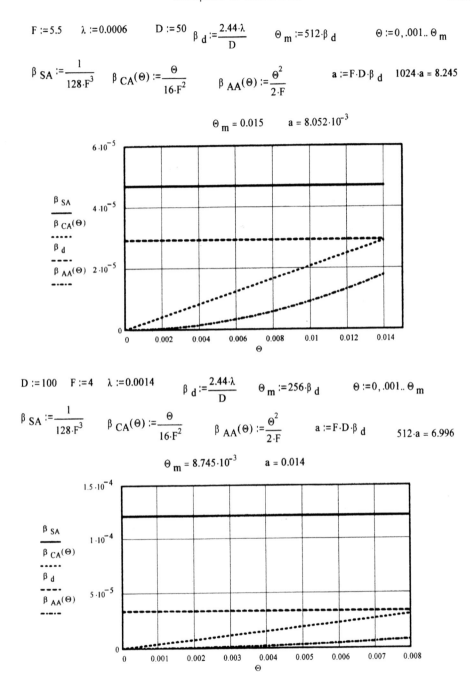

$D := 100$ $F := 4$ $\lambda := 0.0014$ $\beta_d := \dfrac{2.44 \cdot \lambda}{D}$ $\Theta_m := 256 \cdot \beta_d$ $\Theta := 0, .001 .. \Theta_m$

$\beta_{SA} := \dfrac{1}{128 \cdot F^3}$ $\beta_{CA}(\Theta) := \dfrac{\Theta}{16 \cdot F^2}$ $\beta_{AA}(\Theta) := \dfrac{\Theta^2}{2 \cdot F}$ $a := F \cdot D \cdot \beta_d$ $512 \cdot a = 6.996$

$\Theta_m = 8.745 \cdot 10^{-3}$ $a = 0.014$

Figure 5.4. Aberrations of the BOSS.

$c := 2.9975 \cdot 10^{10}$ $c_2 := 1.4388$ $\lambda_1 := 0.0008$ $\lambda_2 := 0.0014$ $T := 300$ $D := 10, 11 .. 100$

$\eta_{sc} := 0.9$ $\tau_a := 0.75$ $\tau_o := 0.95$ $\eta_{ff} := 0.8$ $\eta := 0.8$

$$\beta(D) := \frac{2.44 \cdot \lambda_2}{D}$$

$X := 6.4 \cdot 10^6$ $Y := 9.6 \cdot 10^6$ $R := 33 \cdot 10^6$ $x(D) := R \cdot \beta(D)$

$$N_h(D) := \frac{X}{x(D)}$$

$$N_{fh}(D) := \frac{N_h(D)}{512} \quad N_{fv}(D) := \frac{N_v(D)}{512} \qquad N_v(D) := \frac{Y}{x(D)}$$

$A_{ship} := 30 \cdot 300$ $N_f(D) := N_{fh}(D) \cdot N_{fv}(D)$

$$v := 55 \cdot \frac{1000}{3600} \qquad t_d(D) := \frac{0.1 \cdot x(D)}{v}$$

$$B(D) := \frac{1}{2 \cdot t_d(D)} \qquad k(D) := \frac{A_{ship}}{x(D)^2}$$

$$L_q := \int_{\lambda_1}^{\lambda_2} \frac{2 \cdot c}{\lambda^4 \cdot \left(e^{\frac{c_2}{\lambda \cdot T}} - 1\right)} d\lambda \qquad dL_q := \int_{\lambda_1}^{\lambda_2} \frac{\frac{c_2}{\lambda \cdot T} \cdot e^{\frac{c_2}{\lambda \cdot T}}}{e^{\frac{c_2}{\lambda \cdot T}} - 1} \cdot \frac{1}{T} \cdot \frac{2 \cdot c}{\lambda^4 \cdot \left(e^{\frac{c_2}{\lambda \cdot T}} - 1\right)} d\lambda$$

$$NETD(D) := \frac{\sqrt{B(D)} \cdot \sqrt{L_q}}{k(D) \cdot \lambda_2 \cdot dL_q} \cdot \frac{1}{\sqrt{\eta \cdot \eta_{sc} \cdot \tau_a \cdot \tau_o \cdot \eta_{ff}}}$$

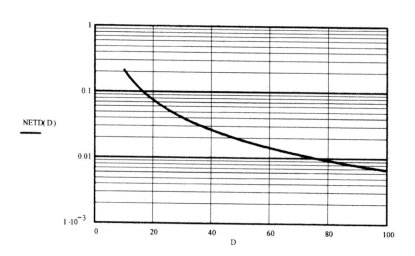

Figure 5.5. NETD in the LWIR of the BOSS.

Another viewpoint can be obtained by considering the weather satellite. It is in a geosynchronous orbit and scans the entire face of the globe in the infrared-- with one detector. There are actually two HgCdTe detectors, but one is for redundancy. They are cooled to about 200 K by a radiative cooler. The scan is accomplished by rotation of the satellite and nodding of a mirror so that a giant spiral is generated in minutes. It doesn't need to go fast because weather doesn't change that fast. This approach could be adopted with a linear array to cover the area more quickly. The resolution would be better than the weather satellite, and the approach is worth considering.

5.6 SECOND ITERATION

It seems clear, as a result of the long-term space application, that an idealized system with photon-limited photon detectors is not the practical way to go. Either moderately cooled photon detectors or thermal detectors should be used. The photon detectors can be cooled either radiatively or with a Peltier cooler; the thermal detectors need not be cooled, but they should be temperature stabilized. Peltier coolers can cool an array to about 200 K at about 1 W. Radiative coolers come in various designs, but in general can cool to about 100 K at 1 W with a size of about 1 m. So we can do HgCdTe in either the MWIR or LWIR.

We have considered the use of microbolometer arrays before, in the design of the night-vision driving system. We considered the existing arrays of 250x350 elements. We can consider that here as well, but we can also anticipate larger formats. In addition, we can work with several arrays in the present format. For instance, two arrays, side by side, will have 500x350 elements. The optics used for the 512x512 system will suffice for this also. A square array of four units subtends 500 by 700 plus a little space. The optics for the 1024x1024 system will provide adequate resolution over the field. This system will improve the NETD by a factor of two. We could, in fact, use a 4x3 array with the 1024 optics and get an improvement of 3.5. The NETD is shown in Fig. 5.6 for an uncooled system. The use of more arrays requires consideration of new optical schemes (an exercise left for the student).

$c_2 := 1.4388$ $c := 2.99795 \cdot 10^{10}$ $h := 6.626 \cdot 10^{-34}$ $\lambda_1 := .0008$ $\lambda_2 := 0.0012$ $\tau_a := 0.8$ $\tau_o := 0.95$

$D := 100$ $\eta_{ff} := .9$ $F := 5.5$ $T := 300$ $\varepsilon := 1$

$X := 6.4 \cdot 10^6$ $Y := 9.6 \cdot 10^6$ $R := 33 \cdot 10^6$ $\alpha := \dfrac{2.44 \cdot \lambda_2}{D}$ $x := R \cdot \alpha$ $x = 966.24$

$N_{fh} := \dfrac{X}{512 \cdot x}$ $N_{vh} := \dfrac{Y}{512 \cdot x}$ $N_f := N_{fh} \cdot N_{vh}$ $N_f = 251.039$

$v := 55 \cdot \dfrac{1000}{3600}$ $t_d := \dfrac{0.1 \cdot x}{N_f \cdot v}$ $B := \dfrac{1}{2 \cdot t_d}$ $B := \dfrac{B}{4}$ $B = 4.962$

$D_{star} := 3 \cdot 10^8, 3.1 \cdot 10^8 .. 9 \cdot 10^9$ $A_{ship} := 30 \cdot 300$ $k := \dfrac{A_{ship}}{x^2}$ $\alpha = 2.928 \cdot 10^{-5}$

$f := F \cdot D$ $a := f \cdot \alpha$ $f = 550$

$a = 0.016$

$$L := \int_{\lambda_1}^{\lambda_2} \frac{2 \cdot c^2 \cdot h}{\lambda^5 \cdot \left(e^{\frac{c_2}{\lambda \cdot T}} - 1\right)} d\lambda$$

$$dL_q := \int_{\lambda_1}^{\lambda_2} \left[\frac{2 \cdot c^2 \cdot h}{\lambda^5 \cdot T \cdot \left(e^{\frac{c_2}{\lambda \cdot T}} - 1\right)^2} \right] \cdot \frac{c_2}{\lambda \cdot T} e^{\frac{c_2}{\lambda \cdot T}} d\lambda$$

$$NETD(D_{star}) := \frac{4}{\pi} \cdot \frac{F}{D} \cdot \frac{\sqrt{B}}{\tau_a \cdot \tau_o \cdot k \cdot \eta_{ff} \varepsilon} \cdot \frac{1}{\alpha} \cdot \frac{1}{D_{star}} \cdot \frac{1}{dL_q}$$

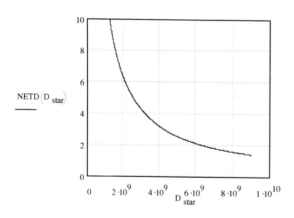

Figure 5.6. The NETD for an uncooled system.

5.7 SUMMARY

We have designed several systems. They are summarized in Table 5.2.

Table 5.2. Characteristics of the BOSS systems.					
Property	Units	PtSi	InSb	HgCdTe	Uncooled
Altitude	Mm	33	33	33	33
Spectrum	μm	3-6	3-6	8-14	8-14
Inst. Field	Mm	0.5	0.25	0.25	0.5
Inst. Field	rad	0.6	0.3	0.3	0.6
Field of Regard	Mm	6.4x9.6	6.4x9.6	6.4x9.6	6.4x9.6
Field of Regard	rad	0.19	0.19	0.19	0.19
Resolution	m	483	483	1127	1127
Resolution	μrad	15	15	34	34
Optics Type	--	Eccentric	Pupil	Paraboloid	Eccentric
Aperture Diameter	cm	50	50	100	100
Mirror Diameter	cm	50	50	100	100
Optical Speed		5.5	5.5	4	4
Focal Length	cm	275	275	400	400
Overall Length	cm	275	275	400	400
Detector Size	μm	41.25	41.25	136	136
Detector Number		1024x1024	512	512	1024x1024
Array Size	mm	42.24	21.12	69.632	136
Scan Rate	fps	1	1	0.5	0.5

5.8 COMMERCIAL APPLICATIONS

One obvious commercial application is the weather satellite. It has provided cloud-cover pictures for many years now and is, perhaps, the way to turn an adversity around. The BOSS, searching for ships at sea, will be limited by cloud cover. A solution is to map it.

Another application is mapping the surface temperature of the oceans, for there is a correlation between areas of warmer water and fish populations. This is the space version of a fish finder. Of course monitoring storm systems such as El Niño could be part of this operation. If such technology were available in the nineteenth century, the Titanic would not have gone down. Iceberg, ocean-liner, and freighter tracking are other good commercial applications.

CHAPTER 6
COLLISION AVOIDANCE REACTION EQUIPMENT (CARE)

Airplanes crash into each other. My colleague, George Zissis, and I were booked on one of the two airplanes that crashed over the Grand Canyon in 1958 before we switched to the redeye. If we had not done that, you wouldn't be reading this! The available air space is getting more crowded, especially near airports. A simple, cheap, effective device that has a high probability of detection and low false alarm rate would be advantageous. This chapter examines how it might be done.

6.1 THE PROBLEM

The challenge is to design an infrared device that can detect the presence of an aircraft at a safe distance throughout the entire sphere that surrounds the host plane. That is easy to specify , but difficult to accomplish.

6.2 GEOMETRY

The field of regard must be the entire sphere that surrounds the aircraft, because collisions can be nose-on, from the sides, top and bottom, or even from the rear. The other craft must be detected anywhere in the sphere. However, it is unrealistic to assume that one sensor can do such a thing. Rather, there must be at least one in the nose and one in the tail, and perhaps one on each wing tip. This is illustrated in Fig. 6.1. Although each sensor is shown as having a 90-degree conical field of view, we might increase this by a few degrees to ensure complete coverage. The geometry includes the safe distance and the size of the aircraft involved. We assume that the planes are traveling at 500 mph. There are different closure rates (with many others that can be interpolated). The head-on case, the most difficult one, involves a closure rate of 1000 mph, i.e., 1600 kmph, 444 m/s. The flank approach is approximately half this, or 222 m/s. The tail chase for these assumptions is not a problem, but it is possible that the host plane has slowed to about 250 mph so that the closing velocity is 111 m/s. If we allow 10 s for reaction time, these distances are 4440 m, 2220 m, and 1110 m. Detection must be accomplished at ranges no less than these. Therefore I choose 5000, 2500, and 1250 meters.

There are several different potential sources of infrared radiation from these aircraft, so the target sizing is a little difficult. Three of the different views to consider are nose-on, side-on, and the rear view.

109

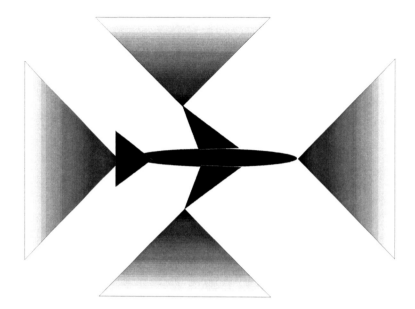

Figure 6.1. Approximate coverage for four avoidance sensors.

The nose-on view includes the front end of the aircraft plus the parts of the jet plume that are not obscured by the body of the plane. It is unlikely that a true nose-on view is ever encountered. The front end of the aircraft consists of the oval cross section of the fuselage plus the front ends of the wings. The representative aircraft can be established by reference to appropriate airliners.[1] A Caravelle is a two-engined 32-m-long craft, has a span of 34.3 m, and a 4-m fuselage diameter. Other aircraft have the following dimensions (in the same order): DC-10--56, 47, and 6; MD-11--61, 47, and 6; L1011--54, 47, and 6; Airbus--54, 45, and 6. These latter are wide bodies, and it would be too optimistic to use their dimensions. So the fuselage can be approximated as circular with a 4-m diameter. The wings can be approximated as 0.3 m thick by about 30 m long. The nose diameter subtends at these three distances about 0.8, 1.6, and 3.2 mrad. A hemisphere contains 2π sr; each sensor covers half of this or π sr; and therefore there are 4.9, 1.2, and 0.3 million pixels in each of these sectors.

The plume is about 120 ft long and 8 ft in circumference (40x3 m^2) . The plume is obscured at dead on, but that dead-on situation almost never occurs (in either aircraft scenarios or darts). We can again assume that the part of the plume that is visible from near head on is 3 m in diameter. It will be shown subsequently that body heat is not a reliable source for detection. The tailpipe cannot be seen from the front, but should not be ignored for the simple tasks of detecting aircraft in front of the subject plane. Figure 6.2 shows the calculations of geometry and dynamics for the different approaches. The reaction time is 10 s, the linear subtense, x, is 4 m, and the field of regard is π sr. For the nose-on situation, the closing velocity is 440 ms^{-1}; the range is the velocity times the reaction time, which turns out to be 4400 m and is reset to 500,000 cm for margin. The angular subtense is x/R; the number of pixels in the field would then be the solid angle divided by the square of the angular subtense, and is almost 5 million, while the angular subtense is 0.8 mr. The other approaches are done similarly.

6.3 SENSITIVITY

A good start is the basic equation (with efficiency factors), and the Earth as a background, because other planes can be below the host plane. This is a detection problem with an unresolved (point) source. We can use Eq. (1.4), here with intensity replacing the product of radiance and area, as shown in Fig. 6.3, but we do not know the radiant intensity. That is why the plot is *as a function* of the radiant intensity. The domain was chosen to include the intensities of the plume, body heat and the tail pipes that are calculated later. We also do not know the bandwidth, because we have not yet chosen the resolution angle nor the scan (or staring) technique. But we must start.

Figure 6.3 shows this start. The source dimension, x, is set at 400 cm, the range at 500,000 cm, the wave band from 0.00043 to 0.00046 cm, the atmospheric transmittance at 0.75, and the optical transmittance at 0.9. Then the angular subtense, α, is calculated, as is the diameter of the optics, D_o, from the diffraction equation. The optical speed is set at the canonical value of 3, and the focal length, f, is calculated. Next the domain of the intensity is established. Then the detector linear dimension, a, the area A_d, and the optical area A_o are all calculated. The bandwidth is calculated based on the time required to cover the field of regard and the number of pixels in that field. The *SNR* is calculated as a function of the intensity of the source, using the previously established Eq. (1.4). The values appear to be quite nice (and not so good) for the intensities we will find–next.

6.3.1. Tailpipe radiation

The several sources of radiation include the tailpipe, the heated body, and the jet plume. The tailpipe radiation can be calculated from data presented in *The Infrared Handbook* (pg 2-81). The tailpipe temperature is 555°C (= 838 K) for the 707B at continuous thrust. Its area (at all times) is 3502 cm^2 and, because it is a pretty good cavity, can be assumed to have an emissivity of one. Then, as shown in Fig. 6.4, the intensity (from both engines) is 309 W/sr and provides an SNR of 3x104, which is wonderful and solves the tail-chase part of the problem.

$$t_r := 10 \qquad x := 4 \qquad \Omega := \pi$$

Nose-on Body

$$v_1 := 440$$

$$R_1 := v_1 \cdot t_r \qquad R_1 = 4.4 \cdot 10^3 \qquad R_1 := 5000$$

$$\alpha_1 := \frac{x}{R_1} \qquad N_1 := \frac{\Omega}{\alpha_1^2} \qquad \alpha_1 = 8 \cdot 10^{-4} \qquad N_1 = 4.909 \cdot 10^6$$

Side-on

$$v_2 := 220$$

$$R_2 := v_2 \cdot t_r$$

$$R_2 = 2.2 \cdot 10^3$$

$$R_2 := 2500$$

$$\alpha_2 := \frac{x}{R_2}$$

$$\alpha_2 = 1.6 \cdot 10^{-3}$$

$$N_2 := \frac{\Omega}{\alpha_2^2}$$

$$N_2 = 1.227 \cdot 10^6$$

Tail-Chase

$$v_3 := 110$$

$$R_3 := v_3 \cdot t_r$$

$$R_3 = 1.1 \cdot 10^3$$

$$R_3 := 1250$$

$$\alpha_3 := \frac{x}{R_3}$$

$$\alpha_3 = 3.2 \cdot 10^{-3}$$

$$N_3 := \frac{\Omega}{\alpha_3^2}$$

$$N_3 = 3.068 \cdot 10^5$$

Figure 6.2. Geometry and dynamics of the CARE.

$x := 400$ $R := 500000$ $\lambda_1 := 0.00043$ $\lambda_2 := 0.00046$ $\tau_a := 0.75$ $\tau_o := .9$

$\alpha := \dfrac{x}{R}$ $D_o := \dfrac{2.44 \cdot \lambda_2}{\alpha}$ $F := 3$ $f := F \cdot D_o$ $I := 0, 100 .. 3000$

$a := f \cdot \alpha$ $A_d := a^2$ $A_o := \dfrac{\pi}{4} \cdot D_o{}^2$ $D_{star} := 10^{10}$ $D_o = 1.403$

$t_r := 10$ $N := \dfrac{\pi}{\alpha^2}$ $t_d := \dfrac{t_r}{N}$ $B := \dfrac{1}{2 \cdot t_d}$

$A_o = 1.546$

$\alpha = 8 \cdot 10^{-4}$

$a = 3.367 \cdot 10^{-3}$

$SNR(I) := \dfrac{\tau_a \cdot \tau_o \cdot A_o}{R^2 \cdot \sqrt{A_d \cdot B}} \cdot I \cdot D_{star}$ $A_d = 1.134 \cdot 10^{-5}$

$f = 4.209$

$SNR(1600) = 40.036$

$SNR(309) = 7.732$ $N = 4.909 \cdot 10^6$

$SNR(0.158) = 3.954 \cdot 10^{-3}$ $B = 2.454 \cdot 10^5$

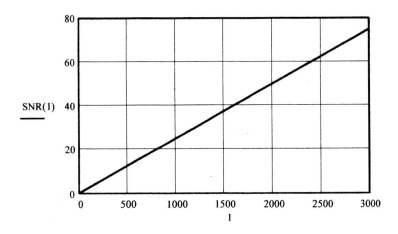

Figure 6.3. General sensitivity calculations.

$$c := 2.99795 \cdot 10^{10} \qquad h := 6.626 \cdot 10^{-34}$$

$$\lambda_1 := 0.00043 \qquad \lambda_2 := 0.00046$$

$$c_2 := 1.4388 \qquad T := 838 \qquad A_s := 2 \cdot 3502$$

$$L := \int_{\lambda_1}^{\lambda_2} \frac{2 \cdot c^2 \cdot h}{\lambda^5 \cdot \left(e^{\frac{c_2}{\lambda \cdot T}} - 1 \right)} d\lambda$$

$$I := L \cdot A_s$$

$$I = 309.102$$

Figure 6.4. The radiant intensity from two 707 jet-engine tailpipes.

6.3.2 Plume radiation

A jet aircraft is propelled by the gasses that are ejected from its tailpipes. These gasses create a plume behind each of the engines that goes back a respectable distance. The gasses are hot, and primarily comprised of carbon dioxide and water vapor.

The pattern of the plume of the engine used on the 707-320B is shown in Fig. 6.5.[2] We can make a reasonable calculation of the total radiant intensity from two engine plumes like this. The calculation is made in Fig. 6.6, and it turns out to be 1.6 kW/sr. Each region has a temperature set for its area and the calculation is in the band for the radiance, and the radiance times the area for the intensity. The assumption was made that the emissivity has the value 1 for the entire pattern. In this narrow spectral region, about 1 in. of atmosphere is completely opaque. In the wings, the emissivity will be about the same because of temperature broadening. The narrow spectral band, 4.3 to 4.6 μm, was chosen because that is the region in which carbon dioxide vapor radiates. This is illustrated in Fig. 6.7, which also

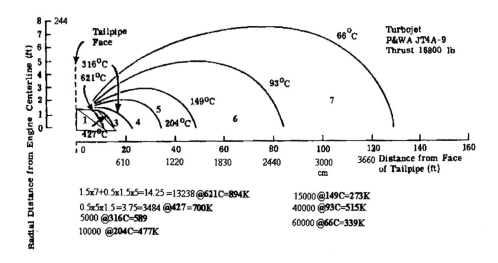

Figure 6.5. Plume radiation from a 707 engine (from *The Infrared Handbook*).

shows that the atmospheric transmission is quite good. In fact, for the ranges considered, it averages 75%. By reference to Fig. 6.3, it can be seen that the SNR is 40. That is more than sufficient, but it was for the entire plume. We will put a little more realism into it in Section 6.3.4.

6.3.3 Body radiation

Aerodynamic friction heats the body of the aircraft. This radiation will be at a much lower temperature than the plume, but there is much more area. This is a highly complicated issue in aerodynamics, and one for which there is relatively little concrete applicable theory. We can make some approximations[2] using the curves in Fig. 6.8. The lower left part of the curve shows that the temperatures are about 500 R (277 K) at sea level for an emissivity of 0.0. A higher emissivity results in a lower temperature. The area is about 30 m by 0.3 m with an emissivity of about 0.1. This results in an intensity of 0.158 and an SNR of about 0.004 (from Fig. 6.3). It is fair to conclude that body radiation resulting from aerodynamic friction is not a significant or dependable radiation source for this application, although it can be quite important for detection of incoming missiles at speeds of about Mach 4.

Although solar heating of the plane might be used, clouds are an issue. The contrast between the plane and the ground or the plane and the sky could be used, but that sounds dicey. Although I might be wrong, I think we must return to the plume radiation.

$$\lambda_1 := 0.00043 \quad \lambda_2 := 0.00046 \quad c := 2.99795 \cdot 10^{10} \quad h := 6.626 \cdot 10^{-34} \quad c_2 := 1.4388$$

$$T := 894 \quad A_s := 13238$$

$$L := \int_{\lambda_1}^{\lambda_2} \frac{2 \cdot c^2 \cdot h}{\lambda^5 \cdot \left(e^{\frac{c_2}{\lambda \cdot T}} - 1 \right)} \, d\lambda \quad I_1 := L \cdot A_s$$

$$I_1 = 748.5$$

$$T := 700 \quad A_s := 427$$

$$L := \int_{\lambda_1}^{\lambda_2} \frac{2 \cdot c^2 \cdot h}{\lambda^5 \cdot \left(e^{\frac{c_2}{\lambda \cdot T}} - 1 \right)} \, d\lambda \quad \begin{array}{l} I_2 := L \cdot A_s \\ I_2 = 8.703 \end{array}$$

$$T := 589 \quad A_s := 5000$$

$$L := \int_{\lambda_1}^{\lambda_2} \frac{2 \cdot c^2 \cdot h}{\lambda^5 \cdot \left(e^{\frac{c_2}{\lambda \cdot T}} - 1 \right)} \, d\lambda \quad \begin{array}{l} I_3 := L \cdot A_s \\ I_3 = 42.417 \end{array}$$

$$T := 477 \quad A_s := 10000$$

$$L := \int_{\lambda_1}^{\lambda_2} \frac{2 \cdot c^2 \cdot h}{\lambda^5 \cdot \left(e^{\frac{c_2}{\lambda \cdot T}} - 1 \right)} \, d\lambda \quad \begin{array}{l} I_4 := L \cdot A_s \\ I_4 = 23.309 \end{array}$$

$$T := 273 \quad A_s := 15000$$

$$L := \int_{\lambda_1}^{\lambda_2} \frac{2 \cdot c^2 \cdot h}{\lambda^5 \cdot \left(e^{\frac{c_2}{\lambda \cdot T}} - 1 \right)} \, d\lambda \quad \begin{array}{l} I_4 := L \cdot A_s \\ I_4 = 0.222 \end{array}$$

$$I := 2 \cdot \left(I_1 + I_2 + I_3 + I_4 \right)$$

$$I = 1.6 \cdot 10^3$$

Figure 6.6. Plume intensity calculations.

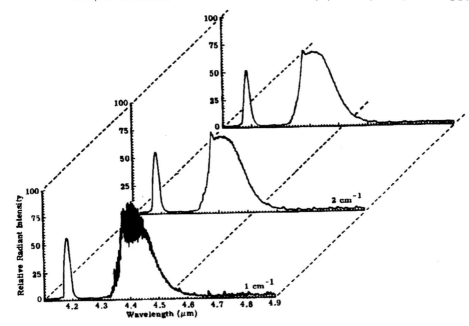

Figure 6.7. Plume radiation through the atmosphere (from pg 24-11 of *The Infrared Handbook*).

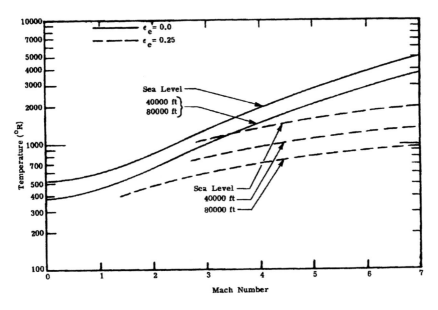

Figure 6.8. Aerodynamic heating, turbulent flat plate, x = 1 ft (from pg 24-11 of *The Infrared Handbook*).

6.3.4 The answer

It seems clear that the solution to this detection problem must be predicated upon radiation from the plume. For this we need to use the very narrow spectral region from 4.35 to 4.65, the SNR must be well above threshold, and the contrast with the background must satisfy the contrast equation

$$k = \frac{I_t A_t + I_n(A - b - A - t) - I_n A_b}{A_b(I_x - I_n)} .$$

(6.1)

This results from requiring that the difference between the flux from a pixel with the target, $I_t\Omega_t$ plus the minimum background, $I_b(\Omega_b - \Omega_t)$ is greater than the difference between the fluxes of the maximum $I_x A_b$, and minimum $I_n A_b$ backgrounds, a tough criterion. The background must be calculated pessimistically on the basis of radiation from the earth. The difference between the radiation from the pixel with the target and without the target is, somewhat surprisingly, a constant that is just equal to the target area times the difference between the target and background radiance, but the contrast (actually the modulation) varies with the size of the pixel. I have estimated (assumed) that the target intensity is 2.5 W/sr--100 times less than the full intensity of two plumes and divided by 2π. This is to account for the obscuration by the plane. Measurements should actually be made of the plume from various forward-aspect angles. The ratio, shown in Fig. 6.9, is gratifying.

6.4 THE OPTICS

The plume and the tailpipes both yield good SNR values. Now the problem is the type of optics to use. It would be nice to have no moving parts. Each system must cover approximately a hemisphere. The half angle of the field is quite large, 0.785 rad, 45 degrees. It should be obvious that a staring optic that would provide the resolution calculated above is just about impossible. We have choices. We can scan. We can make multiple optics. We can broaden the resolution. Note that if we stare, the sensitivity increases significantly. These three choices might be called scanner, fly's eye, and fish eye. They are covered below. And now the time has come to iterate ...iterate...iterate...iterate...iterate...iterate...iterate.

6.4.1 The fly's eye

The fly's eye is an arrangement of a number of staring systems. How many and how big will it get? The diameter required by diffraction limit is 1.1 mm, which is

$\lambda_1 := 0.00043$ $\lambda_2 := 0.00046$ $c := 2.9975 \cdot 10^{10}$ $c_2 := 1.4388$ $h := 6.626 \cdot 10^{-34}$

$T := 290$ $\Omega_t := .4 \cdot 10^{-6}$ $\Omega_b := 10^{-6}$ $L_t := 1, 1.1 .. 2.5$

$$L_n := \int_{\lambda_1}^{\lambda_2} \frac{2 \cdot c^2 \cdot h}{\lambda^4 \cdot \left(e^{\frac{c_2}{\lambda \cdot T}} - 1 \right)} \, d\lambda$$

$T := 310$

$$L_x := \int_{\lambda_1}^{\lambda_2} \frac{2 \cdot c^2 \cdot h}{\lambda^4 \cdot \left(e^{\frac{c_2}{\lambda \cdot T}} - 1 \right)} \, d\lambda$$

$$k(L_t) := \frac{\left[L_t \cdot \Omega_t + L_n \cdot (\Omega_b - \Omega_t) \right] - L_x \cdot \Omega_b}{\Omega_b \cdot (L_x - L_n)}$$

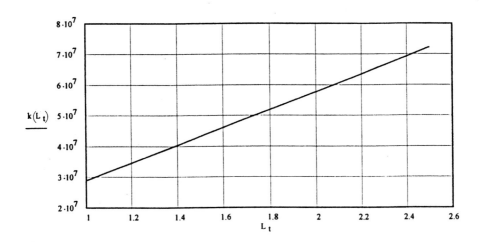

Figure 6.9. Target "contrast."

convenient and small (my mother was right; good things come in small packages–she was about five feet tall). If we use four sensors--nose, tail and wingtips--then each must cover $\pi/2$ radians and the half angle is $\pi/4$ or 45 degrees. There is still too much astigmatism, as Fig. 6.10 shows. In fact, to accommodate this by increasing the F/number would require a speed of about 30. The other approach is to reduce the field so that the astigmatism is acceptable. As shown, this is 0.25 rad half field. Then also, as shown, one needs nine lenses to cover the full field, and nine arrays. This is three lenses on a side that is arranged in a square, octagon, or another good geometry. It is about 5 cm on a side, which is wonderful.

Note that the SNR will be increased over that calculated earlier because the bandwidth is reduced. This is a nice design that has no moving parts but does have nine separate lenses and detectors.

6.4.2 The fish eye

For this design, we disregard resolution and hope that we still have the sensitivity to detect objects in the sector. One lens and one detector will be used. The astigmatism at the edge of the field is 0.103 rad, as shown in Fig. 6.10. This corresponds to a contrast ratio of 2.5, which is a little low.

6.4.3 The scanner

For this we can try to design a scanner with a reasonable number of detectors and scan it over the field of regard. Assume a linear array and a simple optical element. Then scan it at an appropriate rate over the rest of the field. We have just seen in Fig. 6.10 that a half field of 0.25 rad can be accomplished with a resolution of 10 mrad and a simple lens. The full field is obviously twice this, 0.5, and the 90-degree field is 1.57 or about three times this. Thus, we could scan the field thrice with a linear array of 50 elements. The field has to be scanned in ten seconds or less. We can now review the bidding (Fig. 6.11). In summary, the system can consist of a modest array that is at the focus of a single mirror or lens and is scanned over a field of about 90 degrees with a resolution of one mrad. The aperture diameter can be as small as 0.56 cm, the focal length 1.64 cm, and the detector side 34 μm. This system can have a detector with a specific detectivity as low as 10^8 for an SNR of about 250. A modest array of uncooled detectors would probably be sufficient for this.

6.5 PROBABILITY OF DETECTION AND FALSE ALARMS

There is no such thing as certain detection, only a sufficiently high probability of it. False alarms will occur; they must just be infrequent enough that the operator does not ignore the warnings. The subject has been treated exhaustively in the radar

$$\alpha := 0.01 \qquad \Theta := 0, .01 .. \frac{\pi}{4} \qquad\qquad F := 3$$

$$\beta_{SA} := \frac{1}{128 \cdot F^3} \qquad \beta_{CA}(\Theta) := \frac{\Theta}{16 \cdot F^2} \qquad \beta_{AA}(\Theta) := \frac{\Theta^2}{2 \cdot F}$$

$$\beta_{AA}(.78) = 0.101 \qquad \frac{\beta_{AA}(.78)}{\alpha} = 10.14$$

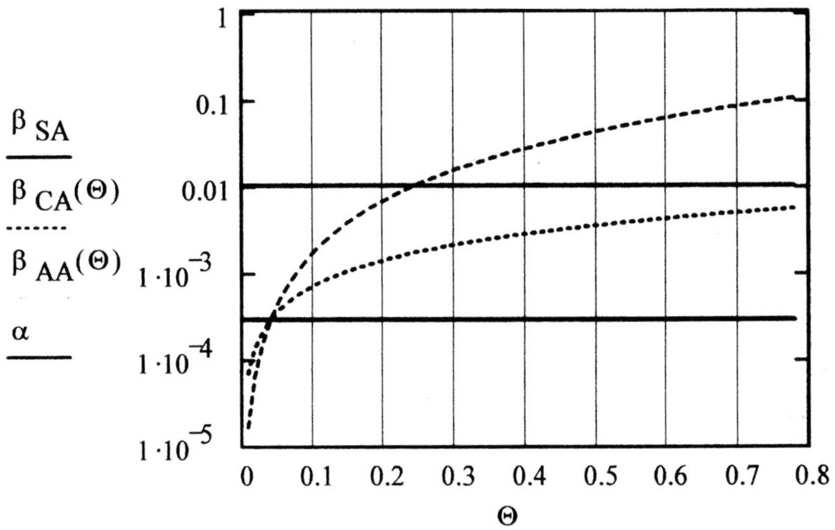

$$\Theta_m := 2 \cdot 0.25 \qquad N := \frac{\pi}{2 \cdot \Theta_m} \qquad N = 3.142$$

Figure 6.10. Optics of the fly's eye.

$$\tau_a := .75 \quad \tau_o := .95 \quad \lambda_2 := 0.00046 \quad \rho := 5000 \quad x := 10 \qquad \alpha := \frac{x}{\rho}$$

$$D_o := \frac{2.44 \cdot \lambda_2}{\alpha} \qquad A_o := \frac{\pi}{4} \cdot D_o^2 \qquad \Theta := \frac{\pi}{4} \qquad F := 3 \quad I := 16.00$$

$$f := F \cdot D_o \quad a := f \cdot \alpha \quad A_d := a^2$$

$$D_{star} := 1 \cdot 10^8, 2 \cdot 10^8 .. 10^9$$

$$m_h := 1 \quad \eta_{sc} := \frac{2}{\pi} \quad m_v := 50 \qquad\qquad f = 1.684$$

$$N_v := \frac{\Theta}{\alpha} \qquad N_h := N_v \qquad\qquad a = 3.367 \cdot 10^{-3}$$

$$\qquad\qquad\qquad\qquad\qquad\qquad \alpha = 2 \cdot 10^{-3}$$

$$B := \frac{N_v \cdot N_h}{2 \cdot t_f m_v \cdot m_h \cdot \eta_{sc}} \qquad SNR(D_{star}) := \frac{\tau_a \cdot \tau_o \cdot A_o}{\rho^2 \cdot \sqrt{A_d \cdot B}} \cdot I \cdot D_{star}$$

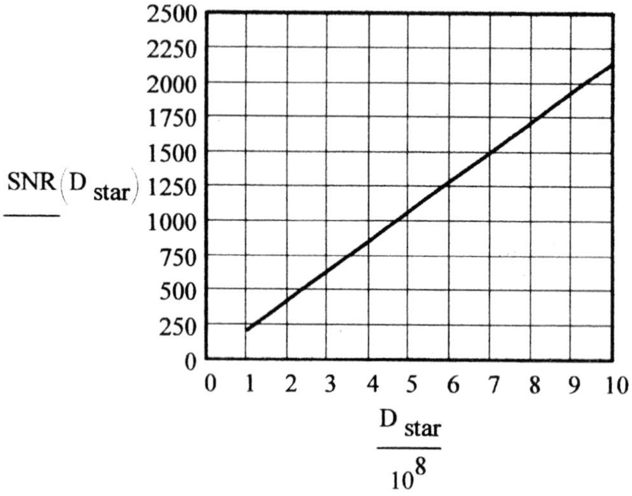

$$I = 16$$

$$B = 242.237$$

$$D_o = 0.561$$

Figure 6.11. Scanner sensitivity calculations.

literature.[3] Figure 6.12 plots the false alarm time (times the bandwidth) as a function of the SNR for different detection probabilities. The bandwidth in this instance is just about 250 and so is the SNR. Thus, for a false alarm only once in a medium-duration flight, say three hours, 10,000 seconds, the log of 10,000 divided times the bandwidth is 6.4. For the SNR of about 250, the detection probability is off the curve, but surely greater than four 9s. I think we have a design. In fact, two: a fly's eye and a scanner. Table 6.1 summarizes the characteristics of the different approaches.

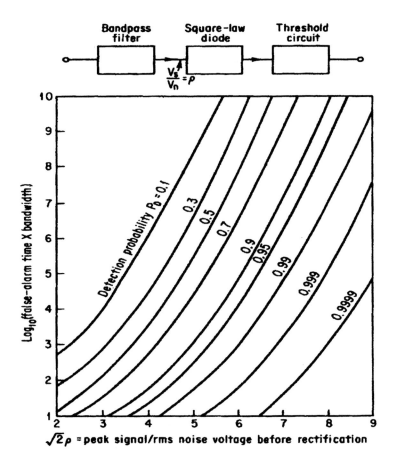

Figure 6.12. Signal-to-noise ratios for detection probability and false-alarm rates (from Jamison[3]).

Table 6.1. Properties of the various designs for each of four units.				
Property	Units	Fly's Eye	Fish Eye	Scanner
Spectrum	μm	0.43-0.46	0.43-0.46	0.43-0.46
Instantaneous Field	rad	0.2	1.57	0.38
Field of Regard	deg	90 cone	90 cone	90 cone
Resolution	m	3	3	3
Resolution	mrad	0.3	0.3	0.3
Range	km	10	10	10
Optics Number		157		1
Optics Type		lens	lens	Eccentric Schwarzschild
Aperture Diameter	cm	2	2	2
Optics Diameter	cm	2	2	2
Optical Speed	F/#	3	3	3
Focal Length	cm	6	6	1.7
Overall Length	cm	6	6	1.7
Detector Size	μm	18	18	5.1
Detector Number	--		1	1
Array Size	μm	18	18	5
Scan Rate	cps	0	0	0.1

The choice between the fly's eye and the scanner is not completely straightforward. One has several lenses; the other has moving parts. We have been able to perform the design procedure. It does seem that this is a viable solution to one of aviation's problems, but such solutions are subject to both economic considerations and preconceptions of the decision makers.

6.6 FINAL THOUGHTS

This was a commercial application. It actually came from the equivalent military use, which is tail warning. Every pilot wants to know as soon as possible if someone has fired a missile at him, mostly from the side or rear. The same considerations of reaction time, range, speed, and detection probability (and false alarms) apply. The air-to-air missiles are smaller and faster and may be detectable by their plumes or body radiation. The latter is worth more consideration than for commercial applications, as the velocities are more like mach 4 or 5 and the body heat is more important. In that case, one may wish to open the spectral band to encompass more than the thermally broadened carbon-dioxide emission.

6.7 REFERENCES

[1]D. Mondey, *The International Encyclopedia of Aviation*, Crescent Books (1988).

[2]W. Wolfe and G. Zissis, *The Infrared Handbook*, The Environmental Research Institute of Michigan (1978); also available from SPIE.

[3]J. Jamison, R. McFee, G. Plass, R. Grube and R. Richards, *Infrared Physics and Engineering*, McGraw Hill (1963).

APPENDIX A
SCANNERS

A.1 INTRODUCTION

As shown in the main body of the text, many infrared systems require devices that scan the instantaneous field of view over a greater field, often called the field of regard. These scanners may be categorized as refractive and reflective and as prismatic and pyramidal.

The pyramidal, prismatic, reflective scanners consist of the plane mirror, a two-sided mirror, a three-sided mirror, and other polygons with other numbers of sides. They are spun about an axis and generate a scan in a single direction. These are shown for four-sided geometries in Fig. A.1. The prismatic scanner, shown on the left, is a geometric polygon that is extended perpendicularly in the third dimension. The pyramidal scanners are pyramidal, although usually the point is blunted.

The refractive scanners are both prismatic and pyramidal, but polygonal in a different way, mainly a pair of prisms.

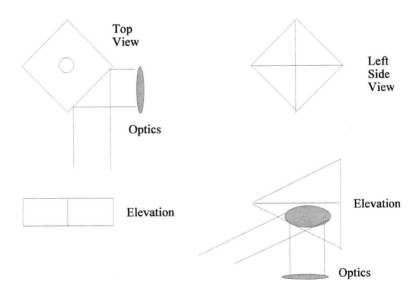

Figure A.1. Prismatic and pyramidal four-sided mirror scanners.

A.2 REFLECTIVE PRISMATIC SCANNERS

Figure A.2 shows a plane mirror used in a prismatic way. The rotational axis is through the center of the mirror face. (The plane mirror used as a monogon pyramid is shown later, but has an axis that is perpendicular to its face). The mirror spins a full 360 degrees, and a portion of the total field is incident on the optical system. The mirror could also be made to oscillate back and forth over the desired field of view. This might also be called a monogon prismatic mirror scanner. The dualgon, a prismatic mirror of two sides, would be this mirror with two reflectorized sides. The trigon is, by extension, a prismatic triangle. The quadrigon, hexagon, and octagon are other popular configurations.[1]

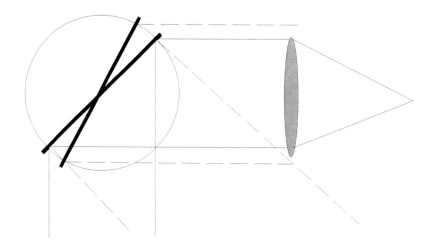

Figure A.2. A single mirror rotating about a facial axis.

The light beam with all of these mirror morphologies exhibits angle doubling; they have a scan efficiency that is less than one and generally have beam wander. They generate a one-dimensional scan unless their facets are offset.

A.2.1 The monogon prismatic mirror

This is a single mirror, reflectorized on one side. As shown in Fig. A.2, this mirror would scan from directly down to an angle off to the right. There is angle doubling, so that as the mirror scans through an angle Θ, the field is swept through an angle 2Θ. At the same time the beam that gets to the entrance pupil is reduced in flux density if the aperture is sized as shown. If it is sized to cover the maximum beam size, there will be not reduction in flux density, but the beam will expand and contract over the entrance pupil, a form of beam wander.

The mirror may rotate through a full 360 degrees, in which case, the scan efficiency is the ratio of the angular size of the field of view to 360 degrees. If the mirror is oscillated resonantly, the scan efficiency is $2/\pi$. This can be verified by calculating the average value of the sine for a positive half cycle,

$$\eta_{sc} = \frac{4}{T}\int_0^{T/4}\sin(2\pi t/T)dt = \frac{4}{T}\frac{T}{2\pi}[\cos(2\pi t/T)]_0^{T/4} = \frac{2}{\pi}. \tag{A.1}$$

The scan excursion can be limited to increase the linearity. Under those conditions the scan efficiency will be reduced.

The mirror may be rotated or oscillated about an axis that is not in its face. In that case the geometry is a little different. It will be covered in the section on quadrigons.

A.2.2 The bigonal plane mirror

This is an ordinary flat mirror that has been reflectorized on both sides. It has twice the scan efficiency as the monogon. When rotated about an axis in its face, it has the same geometric properties.

A.2.3 The trigonal mirror

Just as the single mirror and the two-sided mirror the can cover a total field of view of 180 degrees (and a practical one of much less), the triangular, prismatic scanner can theoretically cover a field of 120 degrees. In general the n-sided prismatic polygon covers a theoretical field of 360/n degrees.

A.2.4 The quadrigon

The four-sided prismatic scanner is shown in Fig. A.3. This covers a theoretical field of 45 degrees. The beam does wander, as described for the plane mirror, but the wander can be eliminated. First, set the top of the optics (shown in this appendix as a simple lens) so that its top limit is where the beam from straight down is the maximum. Then fix the bottom of the lens at the limit of the marginal beam. This can be calculated as follows. As the cube rotates through an angle θ, the tip of the facet moves in x as $r\ sin\ \theta$ and the y coordinate as $r\ cos\ \theta$. The limit of the lens

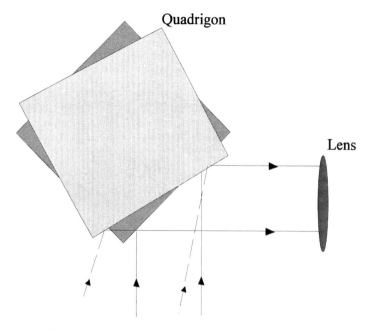

Figure A.3. Rotating prismatic quadrigon.

should them be set at the *y* value for the limiting field angle. This then means that
the facet has an effective facet that is the width times *y*. If the lens is not overfilled,
the diffraction spot will be rectangular, varying in one dimension with the scan
angle. That is, if the width of the facet is *w*, the blur spot will be

$$a = \frac{\lambda}{w} \qquad b = \frac{\lambda}{r\cos\theta} \, . \qquad\qquad (A.2.)$$

However, if the lens is always overfilled (and light is wasted and there is the danger
of double fielding) it is the entrance pupil, and the diffraction blur will be given as
repeated many times in the text. The scan efficiency can be determined to be *y*/*r*.
The active scan can last only from the nadir position to the point at which the tip
crosses the lower line to the lens. Otherwise, light is reflected from two different
directions, a condition I call double fielding.

   The scan efficiency of this scanner is determined by the active scan angle
and the theoretical scan angle. The latter is 360/n, in this case 90 degrees.

A.2.5 The split-aperture scanner[2]

This interesting scanner, shown in Fig. A.4, uses two outrigger mirrors to collect the light from two facets of the rotating scanning element at the same time. The two beams are combined on an optical element after a couple reflections. The scan efficiencies are the same, and the lens should be sized the same as with the quadrigon cube scanner. The next consideration is the placement of the outriggers. Their position determines, in part, the length and width of entire system. The limiting condition can be determined by the left side of the beam adjusted to be just tangent to the circle that is the locus of the cube tips.

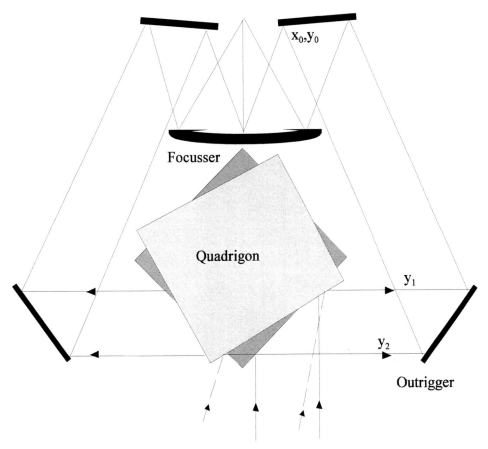

Figure A.4. The split-aperture scanner.

The equation of that line is

$$y = \frac{y_0 - y_2}{x_0} x + y_2 \; .$$

(A.3)

The equation of the circular locus is

$$x^2 + y^2 = r^2 \; .$$

(A.4)

The solution to Eqs. (A.3) and (A.4) is

$$x = \frac{y_0 y_2 + y_2^2}{x_1} \left[1 \pm \sqrt{1 - \frac{4(y_2^2 - r^2)x_1^2}{y_0^2 y_2^2 + y - 2^2)^2}} \right]$$

(A.5)

and

$$y = \left[\frac{y_0^2 - y_2^2}{x_0 x_1} \left(1 \pm \sqrt{1 - \frac{4(y_2^2 - r^2)x_1^2}{y_0^2 y_2^2 + y - 2^2)^2}} \right) + 1 \right] y_2 \; .$$

(A.6)

The values of y_1 and x_2 are set by design; then x_1 and y_2 will be determined by this line. There is a tradeoff between length and width.

An alternative arrangement is to have the "outrigger" mirrors close to the polygon reflect the light to another pair that are angled, as shown in Fig. A.5. This configuration provides essentially normal light to the lens and a minimum breadth to the scanner, but at the cost of extra mirrors and a larger focusing optical element. If the focuser is a mirror, rather than a lens, even more complications arise.

A.2.6 The Weiler wheel

This is a variation on the simple polygon, prismatic scanner. Each of the faces is tilted just a little with respect to the others. Thus, as the wheel spins, a two-dimensional raster is scanned, one line below the other.

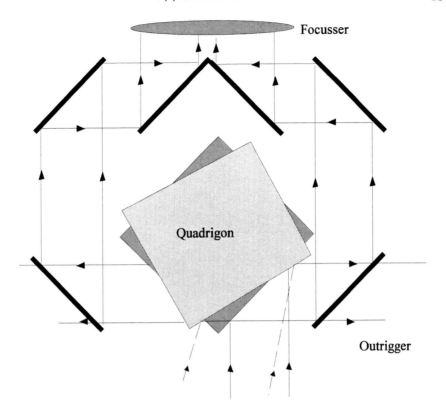

Figure A-5. An alternative split-aperture configuration.

A.3 PYRAMIDAL REFLECTIVE SCANNERS

These devices also start with the monogon, a plane mirror, and include the bigon, a two-sided mirror, the trigon, quadrigon and general n-gons.

A.3.1 The plane mirror

Rather than having the axis of rotation in the face of the mirror or parallel to it, the axis is usually placed at an acute angle with respect to the surface normal. This arrangement is shown in Fig. A.6, which is illustrative of early aircraft-borne strip mappers. The rotation of the mirror moves the field of view into and out of the plane of the paper. As it does so, it also rotates the field of view--at the same rate that the mirror rotates. This monogon can scan a 180 degree field of view. In the example, the axis is at 45 degrees to the normal (and the face). The scan mirror must have a diameter 1.41 times that of the lens.

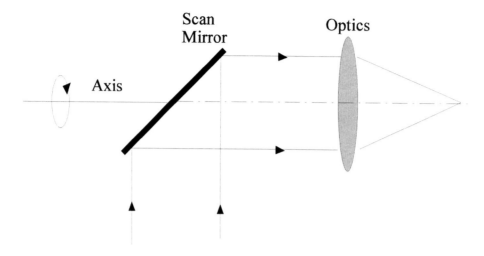

Figure A.6. Single-mirror pyramidal scanner.

The bigon is a plane mirror that is reflectorized on both sides, but it still scans only 180 degrees for a single lens. Some systems used two lenses (optical systems) with a single pyramidal bigon.

A.3.2 The axe blade scanner

This is a bigon in which the two reflecting surfaces are at an angle (greater than zero) to each other, as shown in Fig. A.7. It can be used to scan 180 degrees with a higher scan efficiency than the two-sided mirror. The design of the axe blade includes consideration of the size of the mirror and that of the lens to avoid double fielding. Figure A.8 shows this consideration, and Fig. A.9 shows the scan efficiency as a function of the ratio of the radius of the scanner to that of the lens. It is

$$\eta_{sc} = 1 - \frac{2}{\pi}\arcsin\left(\frac{1}{\dfrac{R}{r}-1}\right) = 1 - \frac{2}{\pi}\arcsin\left(\frac{1}{\rho-1}\right), \qquad (A.7)$$

where R is the scanner radius, r is the lens radius, and ρ is the ratio of R to r.

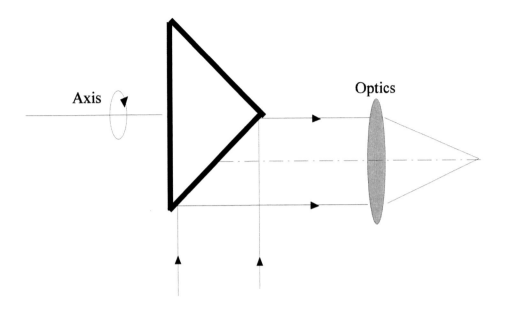

Figure A.7. Axe-blade, pyramidal scanner.

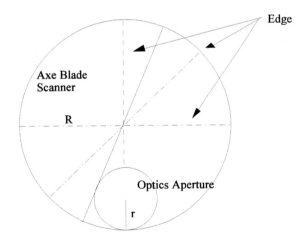

Figure A.8. Limitations in the axe-blade scanner.

$$\rho := 2, 2.1..\ 10 \qquad\qquad \eta_{sc}(\rho) := 1 - \frac{2}{\pi}\,\mathrm{asin}\left(\frac{1}{\rho - 1}\right)$$

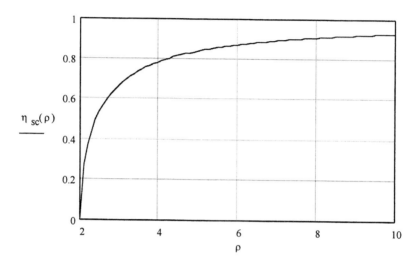

Figure A.9. Scan efficiency as a function of the ratio of scanner radius to the optical aperture.

A.3.3 The pyramid scanner

The geometry of this scanner is that of the Egyptian pyramid, shown in Fig. A.10, a four-sided element that comes to a point. In practice the point is truncated, because it is not used. It is a quadrigon pyramidal scanner. Whereas the axe blade scans 180 degrees with a scan efficiency that approaches 1, the pyramidal scanner covers a field of 90 degrees with a lower range of scan efficiencies. The equation for efficiency is the same, except that a 4 replaces the 2.

The n-gon pyramidal scanner replaces the 2 and 4 with n.

A.4 INSIDE-OUT SCANNERS

The scanners described above may be called external scanners. The incident and reflected light both exist outside the scanner. There is a class of scanners that may be called internal scanners. This geometry is a little difficult to describe specifically and generically, but the concept is intuitively grasped from the figures that follow.

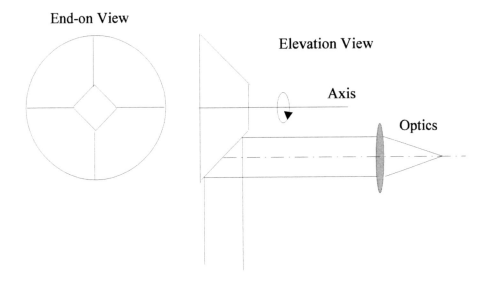

Figure A.10. The pyramidal scanner.

A.4.1 The carousel scanner[3]

Illustrated in Fig. A.11, the carousel consists of an array of mirrors, connected to each other. The light is folded in from out of the plane of the paper to their interior sides in the view shown on the right. As the carousel rotates, the scan lines are made, and either folded back into the plane of the paper or reflected directly to a detector or array. The left part of the figure shows the light entering from the left and reflecting off the folding mirror up to the facet at the top. It then reflects down from the facet to the output fold mirror. Although the figure shows the facet as twice the size of the beam, the design usually has the fold mirrors stick halfway out the sides of the carousel to minimize the required width of the facets. The carousel usually has more than four facets, but four illustrates the principles and is much easier to draw. The right side shows the four facets better and the folding mirror end on. It is a side view of the left part of the figure.

Although it looks like it would be lighter than the equivalent prismatic polygon, the polygon need not be solid, but actually made of the same mirrors with their reflectorized faces on the outside. The scan efficiency is the same, essentially the portion of time the polygon is double fielding. One interesting use of the carousel is to use it both inside and out. Use the inside, for example to scan the field and the outside, with a laser to scan the display.

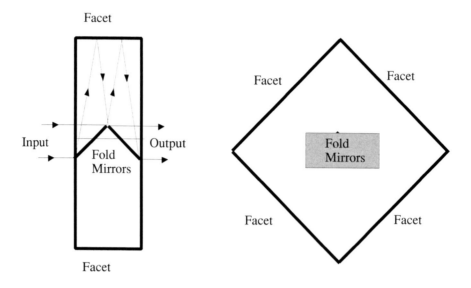

Figure A.11. Carousel scanner.

A.4.2 The soup-bowl scanner[4]

Mike Lloyd describes this scanner that is a form of pyramid scanner. Perhaps it can be considered an inside-out pyramidal scanner; it is shown in Fig. A.12. It has characteristics virtually identical to its outside-out cousin. I have shown only four facets in the bowl. Typically there are more. The scan efficiency is again determined on the basis of eliminating double fielding. It appears that for most applications the outside version is to be preferred, partly because it does not need the folding flat. Because this is a pyramidal scanner, it rotates the field of view.

A.4.3 The saucer scanner[5]

The name the authors used is Compact Thermal Scanner, but to me it resembles a wheel of little saucers, as shown in Fig. A.13.

A.5 REFRACTIVE PRISMATIC SCANNERS[6]

These devices are the refractive analogs of the reflective prismatic scanners. They have the same configurations, but they operate on different principles. Generally,

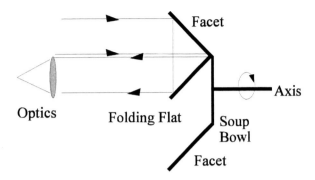

Figure A.12. The soup-bowl scanner.

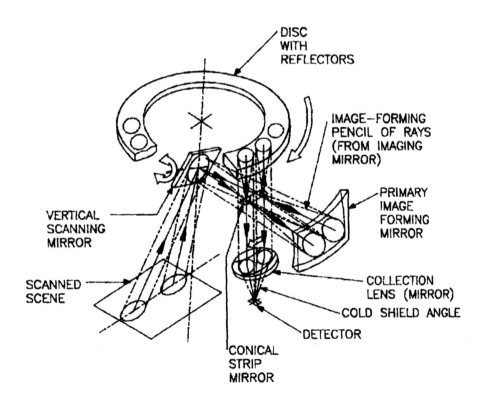

Figure A.13.   The Kollmorgen scanner: an extremely compact real-time scanner. Diamond-turned optics are used throughout the configuration (from Jamieson).[7]

the reflectors deviate the central beam by about 45 degree s and the field is scanned around it; the refractors do not deviate the central beam, which goes straight through. The basis is the operation of the plane parallel plate.

Because these refractive prismatic scanners displace and do not deviate, they must be used in image space where the light converges. When used in object space, they will simply move the beam up and down or back and forth on the entrance pupil--and not change the position of the field of view.

A.5.1 Operation of the plane-parallel plate

Figure A.14 shows how a ray is displaced (but not deviated) by a plane-parallel plate. The ray is deviated by the amount y - y' in Eq. (A.8) below:

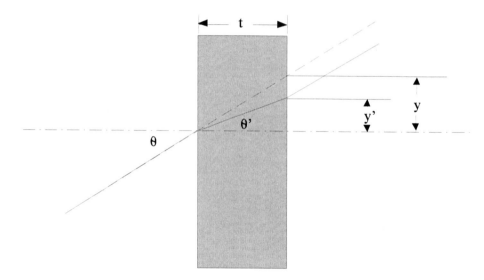

Figure A.14. The geometry of displacement in the plane-parallel plate.

$$\delta y = y - y' = t(\tan\theta - \tan\theta') = t\left(\tan\theta - \tan\left(\arcsin\left(\frac{n}{n'}\sin\theta\right)\right)\right) = t\left[\tan\theta - \sqrt{\left(\frac{n'}{n\sin\theta}\right)^2 - 1}\right]$$

The displacement is a function of the plate thickness and refractive index as well as the angle of incidence. Thus, for a given plate, the displacement changes as the tilt of the plate changes. One can also think of this in terms of the relative refractive index, n'/n, which for most practical matters is the index of the material in air. Since the thickness is set to 1 in the calculations of Fig. A.15, the displacement is relative to the thickness of the plane parallel plate.

$$a := 0, .01 .. \frac{\pi}{3} \qquad n := 1.5 \qquad b(a) := asin\left(\frac{sin(a)}{n}\right) \qquad r_{1.5}(a) := tan(a) - tan(b(a))$$

$$n := 4 \qquad b(a) := asin\left(\frac{sin(a)}{n}\right) \qquad r_{4}(a) := tan(a) - tan(b(a))$$

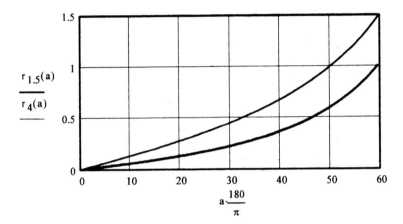

Figure A.15. Displacement of a ray by a plane-parallel plate of refractive index 1.5 and of 4.0.

Because the refractive index is a function of wavelength, the displacement is also a function of wavelength. For a wide range of refractive index values and for different angles of incidence, the change relative to the plate thickness and the refractive index is between 2 and 4. For example, a plate made of germanium that is 10 cm thick and operates in the 3 to 6 µm, the change is 0.048 cm, about 0.5 mm. Germanium is a very low dispersion material.

The plate can scan the entire field, because total internal reflection does not come into play. Note that total internal reflection occurs when

$$n'\sin\theta' = 1 \quad . \tag{A.9}$$

But

$$n'\sin\theta' = n\sin\theta \ .$$
(A.10)

So the condition occurs when $n\sin\theta$ is 1, which is at 90 degrees. The full field is then 180 degrees. It is the geometry of the converging beam and the position of the scanner that determines the usable field of view and the efficiency.

A.5.2 Refractive prismatic polygons

The plane parallel plate is the prototypical prismatic n-gon. If it is thin, it can scan through a full field of 180 degrees, and it works like the bi-gon mirror, that is, in a sense, it is refractive on both sides. Other prismatic polygon separate nicely into those with an even number of sides and those with odd sides (so to speak).

A.6 ROTATING PRISMS

A prism, used in object space, can deviate a beam of light, a ray. Thus a prism can be used in front of the entrance pupil of an optical system to change the direction of the optical axis. If the prism is rotated about an axis through its center and parallel to the surface normal, or coincident with the optical axis, the ray will trace a circular locus. If two prisms are used in sequence, as shown in Fig. A.16, the two circular motions are combined, and a host of different patterns, similar to Lissajous figures can be generated.

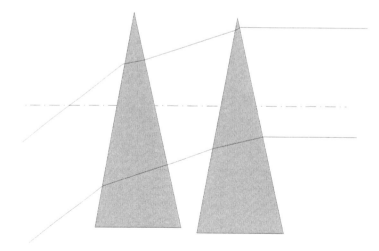

Figure A.16. Geometry of the Risley prism scanner.

The patterns that are generated can be described by a pair of equations that describe the x and y positions of the ray. The basic deviation is given by

$$\delta = (n-1)\sin\alpha \qquad (A.11)$$

where n is the refractive index of the prism and α is its angle. Then,

$$\delta_x = \delta_1\sin(\omega_1 t) + \delta_2\sin(\omega_2 t + \varphi) \qquad (A.12)$$

$$\delta_y = \delta_1\cos(\omega_1 t) + \delta_2\cos(\omega_2 t + \varphi), \qquad (A.13)$$

where ω is the radian frequency of rotation of the prisms, t is time and φ is phase relation between them, that is, the angle between the tips of the prisms at the start of the rotation.

A special arrangement is when the prisms have equal angles and are the same material, so that the deviations are the same. Then, if the frequencies of rotation are the same but the prisms rotate in opposite directions, a line is the locus of the rays. If the prism tips are aligned at the top, then the line will be vertical. If they are aligned horizontally, then the line will be horizontal.

A.7 THE WHEELER LINE

Very often a polygon scanner or a pair of Risley prisms is used as a (fast) line scanner and an oscillating, plane mirror is used to generate frames. These are often and conventionally called the horizontal and vertical scans, respectively. If the framing mirror is oscillated about an axis in its face, the beam wanders over the facets of the polygon, but this beam wander can be reduced to almost nothing by rotating the plane mirror about an offset axis. This was recognized by Bryce Wheeler, who obtained a patent on the technique, actually on the line of axis points.[8] The patent gives the approximate formula for the location of the axes.

A.8 GENERAL OBSERVATIONS

The design of optical-mechanical scanning systems involves spatial visualization, imagination and calculation. The literature[9] is largely in two areas, remote sensing for both military and civilian applications and laser and other printing applications.

In the former, the light is incident upon the scanner from the field and wends its way to the detector (at the speed of light). In printing, the operation is essentially the reverse. There is a source, usually a laser or LED, and its beam is scanned over the field of view. One important piece of advice in scheming and evaluating the remote sensing devices is to reverse the flow of light. Pretend it is a printer; put a source in the position of the detector; and imagine where the light goes. Another piece of advice is to worry about scatter. When attempting to transfer the technology from printing scanners to remote-sensing scanners, remember that most printing scanners use monochromatic light. Some technologies do not transfer, notably holographic and nonlinear techniques.

A.9 REFERENCES

[1] R. B. Barnes, US Patent 3287559 (1966).

[2] H. V. Kennedy, US Patent 32211046 (1965).

[3] M. Weiss, US Patent 3153723 (1964).

[4] J. M. Lloyd, *Thermal Imaging Systems*, Plenum Press (1975).

[5] T. H. Jamieson, *Optical Design of a Compact Thermal Scanner*, Proc. SPIE, **518**, 15, 1984

[6] P. J. Lindstrom, US Patent 3253498

[7] T. H. Jamieson, "Optical design of compact thermal scanner," Proc. SPIE 518, pg 15, Fig. 1 (1984)).

[8] B. Wheeler, US Patent awarded.

[9] L. Beiser, and R. B. Johnson, "Scanners: Chapter 19," in M. Bass, E. Van Stryland, D. Williams and W. Wolfe, eds., *Handbook of Optics*, McGraw Hill (1995); L. Beiser, *Laser Scanning Notebook*, SPIE Press (1992).

APPENDIX B
OPTICAL SYSTEMS

B.1 INTRODUCTION

This appendix provides some details about optical systems that are useful in the infrared. Reference was made in the text to the article by Lloyd Jones,[1] and some of the pertinent data from that article are given here. Both reflective and refractive systems are considered briefly to provide the reader with a comparison to the aberration approximations and a start on representative designs.

B.2 REFLECTIVE OBJECTIVES

Jones lists a number of reflective objectives and gives a table that enumerates their resolution properties. Figures B.1 through B.4 are copies of his figures. This is both an excerpt and an extension of that article. I have chosen just the most pertinent all-reflective objectives that give good wide-angle performance. These are the ones that are pertinent to the applications that have been addressed here. I have put the prescriptions in Zemax and altered the speed to make them more applicable.

The systems discussed include: the SEAL (49), Schwarzschild (19, 21), reflective Schmidt (25), and correctorless Schmidt. The last named was not part of Jones' treatment, but I cited it so often, I thought I needed to include it.

B.1.1 The SEAL

The brief prescription of the F/3 SEAL is:

Element	Radius	Thickness	Conic
1	181.2	-147.8	
2	350.9	147.8	-0.404
stop	infinity	-147.8	
3	350.9	119	-0.404

At 0, 5, 10, and 15 degrees off axis the rms spot diameters are : 0.057, 0.16, 0.43 and 0.88 mrad. That's pretty good, even if the mirrors are large compared to the clear aperture. The sums of the spherical aberration, coma, and astigmatism for 0, 5, 10, and 15 degrees from the approximate equations are: 0.28, 0.89, 1.68, and 3.03. The use of several mirrors and aspheric surfaces was useful.

145

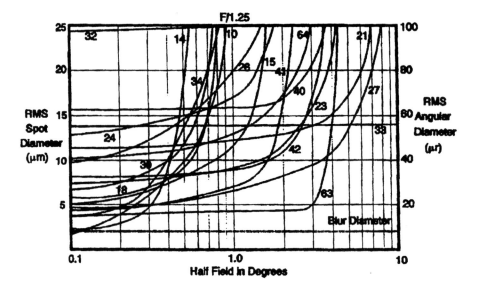

B.1. Catadioptric and reflective objectives; field of view plots are for F/1.25 on a flat image.

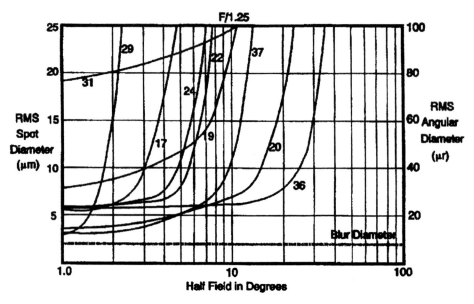

B.2. Catadioptric and reflective objectives; field of view plots are for F/1.25 on a curved image.

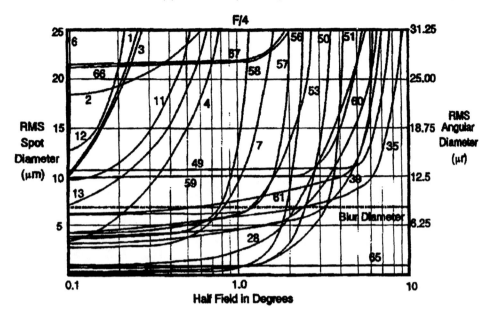

B.3. Catadioptric and reflective objectives; field of view plots are for F/4 on a flat image.

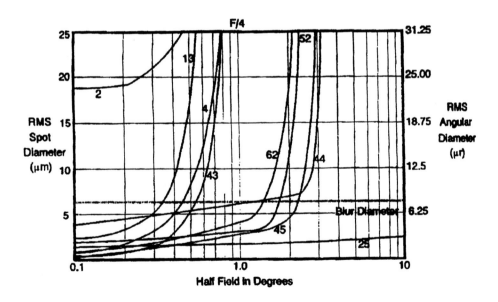

B.4. Catadioptric and reflective objectives; field of view plots are for F/4 on a curved image.

B.2.2 The Schwarzschilds

I have taken the basic design shown by Lloyd, and converted it to an F/3 system and reoptimized. The prescription then is

Element	Radius	Thickness	Conic
1	30.62	-49.44	
2	80.0684	80.26	
stop	infinity	24.62	

The results are that at field angles of 0, 5, 10, and 15 degrees (providing a full field of 30 degrees), one has rms spot sizes of 0, 0.4, 1.62, and 3.7 mrad. Coma is one culprit. When appropriate conic constants are added the results are about 7 percent better. If the image surface has a radius of -27, then the results are spectacular: 0.169 mrad at 15 degrees.

B.2.3 Reflective Schmidt

The prescription given by Lloyd is

Element	Radius	Thickness	Conic
stop	-66752	-67.37	0.508E-7
2	infinity	66.6	
3	-133.97	-66.85	

It results, for an F/3 system, in rms blur diameters of 0.34, 0.19, 1.62, and 4.12 mrad at 0, 5, 10, and 15 degrees. I experimented with the system, that is, used the optimization program in Ken Moore's Zemax. By only adjusting the reflective, corrector plate radius of curvature to -9912.3925, I got 0.92, 0.73, 0.95, and 2.53 mrad at the same angles. One could do better with a higher-order aspheric surface.

B.2.4 Correctorless Schmidt

This is a system that I like very much--if you can afford the space. The downsides are that the overall length must be a little more than twice the focal length, and that, for large fields the image surface is both curved and obscuring. However, the results are nice. Here is the prescription:

Element	Radius	Thickness	Glass	Conic
stop	infinity	0.100		
1	infinity	60		
2	-60	-29.9302	Mirror	
image	-30.0022			

The resolution over a full field is 0.123 mrad. The approximate formula that I used is just about twice this, 0.289 mrad. Notice, however, that I have optimized the focal length and the curvature of the image surface.

The story is not quite as nice for a flat field. The resolution then varies from 1.8 mrad on axis to 2.7 mrad at 15 degrees.

B.3 LENS SYSTEMS

There have been several good treatments of lenses for the infrared, and some designs that provide information about good, wide-angle performance. These are included here. I show what you can do with singlets that are bent for eliminating spherical aberration and minimizing coma; then, the performance of some separated doublets; then, in increasing order of complexity, triplets, mostly of the Cooke variety. Two interesting designs that Brian Bauman did for our helicopter study are also shown. Two of these are for a full field of 25 degrees and one is for 50 degrees.

B.3.1 Singlet lenses

Singlet lenses have two surfaces and a thickness as well as the material as design variables. The lens should be bent to minimize spherical aberration. We will assume that the object is at infinity, as it is for all of the examples and most infrared applications. The position factor is given by

$$p = \frac{i - o}{i + o},$$
(B.1)

so that the position factor for these lenses is -1. The shape factor is given by

$$q = \frac{r_2 - r_1}{r_2 + r_1}.$$
(B.2)

It has been shown that the expression for the shape factor that minimizes spherical aberration is

$$q = -\frac{2(n^2-1)p}{n+2} = \frac{2(n^2-1)}{n+2},$$ (B.3)

where the second expression arises from the fact that p is -1 for this example. Figure B.5 shows this optimum shape factor as a function of refractive index and the ratio of radii for these shape factors. It is quite different in the infrared for materials like germanium, silicon and zinc selenide and zinc sulfide which have refractive indices considerably higher than 1.5. For example, a germanium lens has an index of 4, an optimum shape factor of 5 and a ratio of radii of 2/3. Lenses of different speeds are described below, and are compared to the third-order approximations used so often in this text.

In the Zemax file is a singlet lens made of ZnSe. It has radii of 311.4638 and 755.5295, thickness of 20 mm and an optical speed of F/3.21. Its performance is summarized in Table B.1. A similar germanium singlet has radii of and , a thickness of and is an F/3.02 lens. Its performance is listed in the third column.

The fourth column lists the sum of spherical, coma and astigmatism as calculated by the formulas. Note that the Airy disk diameter for the maximum wavelength for this lens is 0.29 mrad. Values better than this do not indicate that the lens is better than diffraction limit. They are ray traces.

Table B.1 Performance of two singlet lenses for the 8-12 μm region.			
Field Angle [degrees]	Resolution [mrad] ZnSe	Resolution [mrad] Ge	Formula Resolution [mrad]
0	0.076	0.072	0.565
1	0.036	0.080	0.637
2	0.125	0.200	0.803
3	0.360	0.432	1.065
4	0.700	0.760	1.421
5	1.120	1.187	1.872
10	4.800	4.84	5.551
15	25.00	11.3	12.000

$$n := 1, 1.1 .. 5$$

$$q(n) := \frac{2 \cdot \left(n^2 - 1\right)}{n + 2}$$

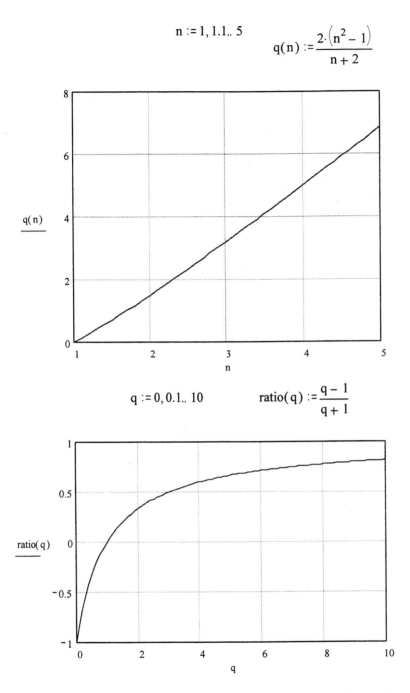

$$q := 0, 0.1 .. 10$$

$$ratio(q) := \frac{q - 1}{q + 1}$$

Figure B.5. The shape factor as a function of refractive index and ratio of radii as a function of shape factor.

These are two examples of singlets for the infrared. A complete compendium, even for singlets, is out of the question. First, they can vary in spectral range by a little for the LWIR. They can be of different speeds and materials. In fact, the results vary some even when optimization is done for different fields of view. This can only be a taste.

B.3.2 Doublet Lenses

These can be either cemented, i.e., with no spacing between the two elements, or separated. The cemented doublet can be made achromatic using two different materials. For the achromatic, separated (including contact) doublet the chromatic aberration is zero if

$$\phi_A = \frac{V_A}{V_A - V_B}\phi \qquad \phi_B = \frac{V_B}{V_B - V_A}\phi \,, \qquad (B.4)$$

where ϕ is the lens power and V the dispersion for lenses A and B. Obviously for this we need a table of V numbers (Abbe numbers) for the different materials. This is Table B.2 for a selected number of infrared lens materials.

Table B.2. Abbe numbers for selected infrared lens materials.		
	3-5 μm	8-10 μm
Germanium	100	1200
Silicon	250	900
Zinc Selenide	190	80
Zinc Sulfide	150	80
Amtir I	200	150
Arsenic Trisulfide	180	50

These data are approximate because one usually has a very specific spectrum, and it will not be exactly 3 to 5 or 8 to 10, or any other specific numbers. Look them up in your Wolfe and Zissis. The refractive index values are all there, and the definition of the V number for the infrared is

$$V = \frac{n_{short} - n_{long}}{n_{middle} - 1} .$$ (B.5)

In the visible these are the F, C, and d lines of the Fraunhofer spectrum. In the infrared they can be all sorts of spectral bands. They can be 3 to 5 µm, 3.1 t0 4.8 µm, 7.6 to 10.2 µm, etc. So, not only are there about eight different materials (Ge, Si, ZnS, ZnSe, Amtir, $CaAl_2O_3$ glass, AsS_3 glass and sometimes a few others), there are also three surfaces and two thicknesses. Even the reasonable combinations are in unreasonably large number. Contact doublets have four radii. Four examples from Bob Fischer's lecture "Optical Design for the Infrared" are given below; the prescriptions in Table B.3 and the results in Table B.4. They are all 1-mm-diameter F/1.5 lenses for the LWIR. I optimized the back focal length and got trivial differences from Bob. I evaluated for 8-12µm; he gave ray fans for 8-10µm.

Table B.3 Doublet prescriptions.				
Lens	Amtir-1/ZnS	Ge/Amtir-1	Ge/ZnS	Ge/ZnSe
R1	1.3874	1.5184	1.6377	1.6572
R2	5.1990	2.2313	2.4599	2.5151
R3	-11.5083	-98.7706	-1.4296	-1.3081
R4	19.8665	31.8018	-1.4730	-1.3562
t1	0.1724	0.1340	0.1319	0.1319
t2	0.0327	0.0532	0.1307	0.1380
t3	0.1	0.1	0.1	0.1
bfl	1.2960	1.2984	1.2743	1.2793

Table B.4. The performance of some infrared doublet lenses.				
Lens	Amtir-1/ZnS	Ge/Amtir-1	Ge/ZnS	Ge/ZnSe
Field Angle [degrees]	Resolution [mrad]	Resolution [mrad]	Resolution [mrad]	Resolution [mrad]
0	0.40	0.18	0.11	0.46
1	0.40	0.11	0.13	0.45
2	0.53	0.24	0.34	0.49
3	1.50	0.72	0.76	0.77
4	1.60	1.39	1.35	1.28
5	2.50	2.27	2.13	2.00
10	10.3	9.83	8.73	8.29
15	24.4	23.39	20.27	19.25

These lenses all perform about the same. The first does seem to be worst. The others distribute the resolution differently. They are better than the simple equations predict. Some good design (by Fischer) was done. It is also true that the main aberration at the larger angles is curvature of field.

B.3.3 The (Cooke) triplet

This triplet has three air-spaced lenses that are used to both color correct and balance other aberrations. I tried two designs, one was germanium and zinc selenide; the other replaced the zinc selenide with silicon. The form came out different from the visible, Cooke, triplet that has two biconvex and one biconcave lens. Mine has three menisci. The optimum shown by Fischer has two biconcave germaniums surrounding a biconvex. (A good process with a modern design program is to vary a limited number of parameters at a time. If you vary all six radii and all six spacing *ab initio*, you are very likely to get garbage. However, if you vary things one or a few at a time you are likely to drive the system to a local minimum. The way out with Zemax is to use the hammer optimization.) The triplet looks like this:

Element	Radius	Thickness	Glass	Conic
Stop	18.3021	0.9499	germanium	
1	15.3831	1.0000		
2	19.9701	1.0072	zinc selenide	
3	18.5201	1.0000		
4	17.6411	1.1149	germanium	
5	30.3345	20.2265		

The resulting resolution values for 0, 1, 2, 3, 4, 5, 10, and 15 degrees of field angle were 0, 0.056, 0.213, 0.476, 0.847, 1.326, 5.446 and 12.9 milliradians. The silicon lens was three percent better at the last two angles. I could probably obtain a little more performance out of this by moving the stop and trying some aspherics, but that is not the purpose of this text..

If one forces the triplet to be of the Fischer optimum form, the results are better. For the same angles the rms spots diameters are, 0.256, 0.256, 0.258, 0.262, 0.268, 0.278, 0.452, and 1.06 mrad. The prescription looks like this:

Element	Radius	Thickness	Glass	Conic
Stop	-18.2818	2	germanium	
1	-23.2325	2		
2	116.8642	2	zinc selenide	
3	89.0374	2		
4	-107.8522	2	germanium	
5	-42.3473	2		

It should be obvious that I forced all thicknesses to be 2. A better system results if one allows the lenses to separate, the second and third lenses move away from the first to give a still better performance, approximately 0.1 mrad our to 5 degrees then 0.17 at 10 and 0.32 at 15. One can also force the three lenses to be in contact (at least at the edges) to obtain a system that is almost as good.

B.3.4 Fifty-degree cold helicopter lens.

It was shown in the discussion of the helicopter infrared viewer that one would like a high resolution field of view of a full 50 degrees. A lens for this is illustrated in Fig. B.6 and its MTF is listed in Fig. B.7. Its prescription is:

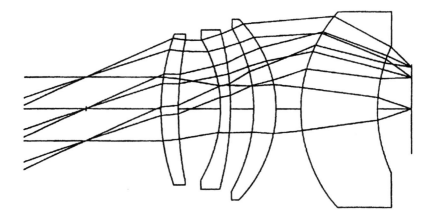

Figure B.6. the fifty-degree cold lens (F/1.5   +/- 25).

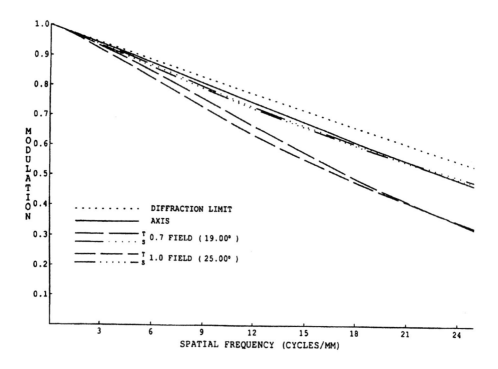

Figure B.7. MTFs of the fifty-degree lens.

Element	Radius	Thickness	Material	
object	infinity	infinity		
stop	infinity	30.395511		
2	9670241	7.050114	Ge	
3	155.84290	17.455786		
4	-57.92156	4.0000000	ZnSe	
5	-129.94860	9.406118		
6	-68.64464	9.000000	ZnSe	
7	-54.34028	10.230996		
8	61.94634	31.449508	Ge	
9	69.09355	13.701141		0.743390
image	infinity			

B.4 REFERENCES

[1]Jones, L., Chapter 18 in M.Bass, E. Van Stryland, D. Williams and W. Wolfe, *Handbook of Optics*, McGraw Hill, 1995.

APPENDIX C
ORBITAL VELOCITIES AND TIME

Orbital velocities and periods can be calculated by a more accurate formula than that used in the text. The formula is[1]

$$v = \sqrt{2gR_E^2\left(\frac{1}{R} - \frac{1}{R_{ave}}\right)} \,,$$

(C.1)

where v is the orbital velocity, g is the gravitational force, R_E is the radius of the Earth, R is the instantaneous satellite height, measured from the center of the earth, R_{ave} is the average radius of the orbit, also measured from the center of the Earth.

The orbital period is given by

$$T = \frac{2\pi a^{3/2}}{R_E\sqrt{g}} \,.$$

(C.2)

The calculation is made in Fig. C-1.

The velocity at orbital altitude is a function of the altitudes of apogee and perigee, with the subscripts ap and per, as measured from the center of the earth. That is why the altitudes from the surface have added to them the radius of the Earth. The velocity one needs for the scanner calculations is the linear velocity on the surface of the Earth; it is different from that at altitude. The angular velocity is constant, so the two velocities are related by the ratio of radii. Notice that for a 200 km altitude, the velocity is about 7.6 km/s, very close to the approximation made in the main part of the text.

---

[1]J. L. Meriam, *Dynamics*, p.179.

$$h_{ap} := 500000 \cdot m \qquad h_{per} := 100000 \cdot m \qquad R_E := 6378245 \cdot m \qquad g = 9.807 \frac{m}{s^2}$$

$$R_{ap} := h_{ap} + R_E \qquad R_{per} := h_{per} + R_E \qquad h := 100000, 300000 .. 1000000$$

$$R_{ave} := \frac{R_{ap} + R_{per}}{2} \qquad R(h) := h + R_E$$

$$v(h) := \sqrt{\left(2 \cdot g \cdot R_E^2\right) \cdot \left(\frac{1}{h + R_E} - \frac{1}{2 \cdot R_{ave}}\right)} \qquad T := \frac{2 \cdot \pi \cdot R_{ave}^{\frac{3}{2}}}{R_E \cdot \sqrt{g}}$$

$$v_E(h) := v(h) \cdot \frac{R_E}{R(h)} \qquad\qquad T = 5.429 \cdot 10^3 \ s$$

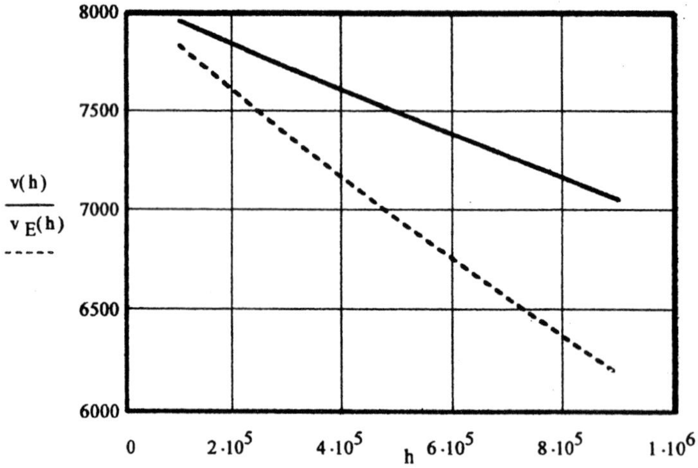

Figure C.1. Orbital velocities at altitude and on the Earth.

INDEX

161

WILLIAM L. WOLFE was born in Yonkers, New York, at a very early age. He received his BS in physics from Buckness University, cum laude. He did graduate work at the University of Michigan, where he received an MS in physics and an MSE in electrical engineering. He held positions of Research Engineer and lecturer at the University of Michigan. He was later Chief Engineer and Department Manager at the Radiation Center of Honeywell, in Lexington, Massachusetts. In 1969 he became Professor of Optical Sciences at the University of Arizona Optical Sciences Center, where he taught infrared techniques and radiometry. In 1995 he became Professor Emeritus. He has been a fellow and on the board of directors of the Optical Society of America; a senior member of the IEEE; and a fellow, life member, and past president of SPIE–The International Society for Optical Engineering. He is the coeditor of *The Infrared Handbook,* and associate editor of the second edition of the *Handbook of Optics.* He has recently written three other Tutorial Texts: *Fundamentals of Infrared system Design, Fundamentals of Imaging Spectrometers* and *Fundamentals of Radiometry.* He was awarded the Gold Medal of the Society in 1999. Present activities include the investigation of cancer detection by optical techniques, colposcopy and more writing. He is the proud father of three wonderful children and five grandchildren (including twins). In his spare time, which is very limited because he is retired, he fly fishes, sings, gardens and uses his wife's telephone and computer.